VIBRATIONAL SPECTROSCOPY
FROM THEORY TO PRACTICE

Vibrational spectroscopy. From theory to practice
Editor Kamilla Malek

AUTHORS

Barańska Małgorzata, Prof., PhD, DSc	Faculty of Chemistry, Jagiellonian University in Krakow
Bukowska Jolanta, Prof., PhD, DSc	Faculty of Chemistry, University of Warsaw
Chmura-Skirlińska Antonina, PhD	Jagiellonian Centre for Experimental Therapeutics (JCET), Jagiellonian University in Krakow
Chruszcz-Lipska Katarzyna, PhD	AGH University of Science and Technology, Faculty of Drilling, Oil and Gas
Czamara Krzysztof, MSc	Faculty of Chemistry, Jagiellonian University in Krakow
Dybaś Jakub, MSc	Faculty of Chemistry, Jagiellonian University in Krakow
Gąsior-Głogowska Marlena, PhD, Eng.	Faculty of Fundamental Problems of Technology, Wroclaw University of Science and Technology
Jaworska Aleksandra, PhD	Faculty of Chemistry, Jagiellonian University in Krakow
Kaczor Agnieszka, PhD, DSc	Faculty of Chemistry, Jagiellonian University in Krakow
Kochan Kamila, MSc	Faculty of Chemistry, Jagiellonian University in Krakow
Królikowska Agata, PhD	Faculty of Chemistry, University of Warsaw
Lipiński Piotr, F.J., PhD	Mossakowski Medical Research Centre, Polish Academy of Sciences
Majzner Katarzyna, PhD	Faculty of Chemistry, Jagiellonian University in Krakow
Malek Kamilla, PhD, DSc	Faculty of Chemistry, Jagiellonian University in Krakow
Marzec Katarzyna M., PhD	Jagiellonian Centre for Experimental Therapeutics (JCET), Jagiellonian University in Krakow
Miśkowiec Paweł, PhD	Faculty of Chemistry, Jagiellonian University in Krakow
Oleszko Adam, MSc, Eng.	Faculty of Fundamental Problems of Technology, Wroclaw University of Science and Technology
Pacia Marta Z., MSc	Faculty of Chemistry, Jagiellonian University in Krakow
Rode Joanna, PhD	1. Institute of Nuclear Chemistry and Technology 2. Institute of Organic Chemistry, Polish Academy of Sciences
Ryguła Anna, PhD	Jagiellonian Centre for Experimental Therapeutics (JCET), Faculty of Chemistry, Jagiellonian University in Krakow
Staniszewska-Ślęzak Emilia, MSc	Faculty of Chemistry, Jagiellonian University in Krakow
Wiercigroch Ewelina, MSc	Faculty of Chemistry, Jagiellonian University in Krakow
Wróbel Tomasz, PhD	Faculty of Chemistry, Jagiellonian University in Krakow
Zając Grzegorz, MSc	Faculty of Chemistry, Jagiellonian University in Krakow

VIBRATIONAL SPECTROSCOPY

FROM THEORY TO PRACTICE

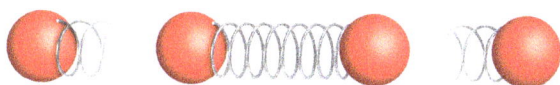

edited by
Kamilla Malek

PWN

Cover and title page design:
Marek Goebel

Project editor:
Katarzyna Włodarczyk-Gil

Editorial coordinator:
Iwona Lewandowska

Copy editor:
Łukasz Boda

Production coordinator:
Mariola Grzywacka

Edited by:
Kamilla Malek, Faculty of Chemistry, Jagiellonian University

DTP:
Dariusz Ziach

The publication was co-funded by the Faculty of Chemistry
of the Jagiellonian University in Krakow

ISBN 978-83-01-18978-5

First edition

Wydawnictwo Naukowe PWN SA
02-460 Warszawa, ul. Gottlieba Daimlera 2
tel. 22 69 54 321, faks 22 69 54 288
infolinia 801 33 33 88
e-mail: pwn@pwn.com.pl; reklama@pwn.pl

Content

Fundamentals of infrared spectroscopy (FT-IR)

Kamilla Malek, Emilia Staniszewska-Ślęzak, Kamila Kochan, Katarzyna Majzner

The interaction of matter with electromagnetic radiation is fundamental in optical spectroscopy and it relies on absorption, emission and scattering of light photons. Due to such interactions one registers a spectrum consisting of bands of a given frequency, shape and intensity. As the title of this Chapter indicates infrared absorption spectroscopy results from absorption of light quantum from IR region, which matches to energy difference between vibrational levels. The IR spectrum is usually presented in a scale of wavenumbers [cm^{-1}] and can be collected in three regions of IR radiation. It should be emphasized that IR and Raman techniques (described in Chapter 2) are complementary to each other, as both relay on registration of transitions between vibrational levels, but information that they provide is complementary to each other. This can be illustrated by the rule of mutual exclusion, which states, that if a molecule has a center of symmetry, vibration active in infrared absorption spectrum is inactive in Raman and *vice versa*. The most frequently used IR region is mid-infrared radiation (MIR, *Mid InfraRed*), which covers the region of 400 – 4000 cm^{-1}. Within this IR spectrum, a region from 400 to 1500 cm^{-1}, called "fingerprint", is specific for each chemical species and this fact strictly results from characteristics of vibrational motion described by normal modes. A region of far infrared radiation (FIR, *Far InfraRed*, 50 – 400 cm^{-1}) also registers normal modes while a region above 4000 cm^{-1} (NIR, *Near InfraRed*) provides information about overtones and combination modes. The abbreviation *FT* in the name of the technique means *Fourier transform* – the mathematical operation that decomposes a function of time (a *signal*) into the function of frequency. FT-IR spectroscopy is applied for the identification and quantitative analysis of chemical compounds or their mixtures as well as the determination of physicochemical features such as molecular structure and its changes due to a stress/reaction, kinetics of reactions and intramolecular dynamics. IR spectra of a sample in any state can be collected.

1.1. Quantum description – models of the harmonic and anharmonic oscillators

Molecules consisting of nuclei and electrons, have a certain reserve of internal energy, which manifests itself in various forms of movement, e.g. vibrations. A molecule containing N atoms can perform $3N - 6$ vibrations or $3N - 5$, if it is linear. An exceptional feature of oscillations is non-zero value of their potential energy in the ground vibrational state, therefore vibrations of molecules take place in every phase and do not stop even at temperature of 0 K. As any energy in the micro-world, vibrational energy is also quantized, what means that its values are limited to certain discrete values called vibrational levels. Therefore, each chemical species adopts or donates energy only in characteristic portions called quanta. The energy of discrete vibrational levels is determined by solving the Schrödinger equation for a model of a harmonic oscillator. The eigenvalues of the vibrational Schrödinger equation for diatomic molecule are given by (Eq. 1.1):

$$E_{\text{osc}} = h\nu \left(\upsilon + \tfrac{1}{2}\right), \tag{1.1}$$

where: ν is frequency of the harmonic oscillator $\nu = \frac{1}{2\pi}\sqrt{\frac{f}{\mu_{\text{red}}}}$, f – force constant, μ_{red} – reduced mass $\mu_{\text{red}} = \frac{m_1 m_2}{m_1 + m_2}$, and υ – a vibrational quantum number ($\upsilon = 0, 1, 2, 3, \ldots$).

The value of the vibrational quantum number in Eq. 1.1 determines quantum nature of vibrational motion and its energy also depends on the type of the molecule since this expression includes the value of force constant and reduced mass of the chemical species.

During oscillations, potential energy reaches the maximum value at the largest deflection from the equilibrium, the potential energy curve change is illustrated by a parabola (Figure 1.1).

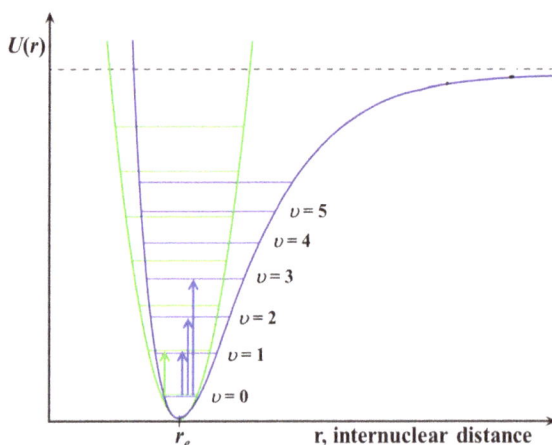

Fig. 1.1. Potential energy U curve for the models of the harmonic oscillator (green line) and of the anharmonic oscillator (blue line). Vertical arrows label allowed transitions between vibrational levels υ

Potential energy curve in Figure 1.1. shows that energy of the harmonic oscillator never reaches zero value, since the energy is ½hv for the lowest allowed vibrational level $\upsilon = 0$. The vibrational levels of the harmonic oscillator are equally separated by hv and therefore energy difference does not depend on the vibrational quantum number υ and is affected by chemical species only. Vibrational spectrum of the harmonic oscillator exhibits only one band since it represents transition for a diatomic molecule. A required condition to record the IR absorption spectrum is a change of dipole moment of a molecule during vibration, wherein the change of the vibrational level can be only $\Delta\upsilon = 1$ (a green arrow in Fig. 1.1).

The transition moment of absorption must be non-zero and is given by (1.2):

$$\mu_{nm} = \langle \Psi_n | \hat{\mu} | \psi_m \rangle \neq 0, \tag{1.2}$$

and its operator is the change of the dipole moment $\hat{\mu}$ due to transition from n to m vibrational level. This transition is called a fundamental tone. Due to the fact that the most occupied vibrational level at room temperature (according to the Boltzmann energy distribution) is $\upsilon = 0$, bands observed in IR spectrum mainly originate from transitions from $\upsilon = 0$ to $\upsilon = 1$ (the green arrow in Fig. 1.1).

The model of the harmonic oscillator is an idealized model, whereas in fact, oscillations are anharmonic, as the Hooke's law is not satisfied. In this case the potential energy curve is illustrated by the Morse curve (Fig. 1.1.B) and energy of the anharmonic oscillator is expressed by an equation 1.3:

$$E_{osc}^{anh} = hv\left(v + \frac{1}{2}\right) - hvx\left(v + \frac{1}{2}\right)^2, \tag{1.3}$$

while neighboring vibrational levels are separated by (1.4):

$$\Delta E_{osc}^{anh} = E_{v+1} - E_v = hv[1 - 2x(v + 1)], \tag{1.4}$$

where the parameter x is an anharmonic constant that describes deviation of anharmonic oscillator from the ideal harmonic model. It takes considerable values for vibrations of a high energy.

It is clear from Figure 1.1.B and equation (1.4) that the intervals between vibrational levels are not constant, but they decrease along with increasing quantum number υ. The Morse potential also shows dissociation of a bond which appears for large values of υ. In addition, the model of the anharmonic oscillator extends the selection rules. Transitions become allowed not only for $\Delta\upsilon = 1$ (fundamental tones) but also for $\Delta\upsilon = 2, 3 , ...,$ sequentially called the first, second (etc.) overtone (blue arrows in Fig. 1.1). Their intensities are much lower than intensity of the fundamental tone.

1.2. Normal coordinates

In general, a molecule built from N atoms has $3N - 6(5)$ internal degrees of freedom. They describe oscillations, for which a concept of the normal coordinates/

vibrations/modes was introduced. A normal coordinate describes a simultaneous movement of all atoms, occurring with the same frequency and consistent in phase, although amplitude of vibration of individual atoms can be different. Vibrations do not induce translation or rotation of the molecule. Individual normal modes are orthogonal to each other, which means that their movement is mutually independent.

Extremely helpful in the analysis of IR and Raman spectra is to consider symmetry of a molecule (and therefore – its possible molecular structure) as well as its influence on spectral features. $3N - 6(5)$ normal modes of N-atomic molecules can be classified according to properties of a point group. For example, the molecule of H_2O, which belongs to the C_{2v} group, has two vibrations with A_1 and one with B_2 symmetry. The reader can get familiar with details of group theory in the literature [1].

It should be also emphasized that a number of vibrations active in IR and Raman spectra and predicted on the basis of group theory can be different from a number of bands observed in experimental spectrum. These discrepancies can result from the presence of overtones and combinatorial bands, Fermi resonance, the abolition of the degeneracy of some vibrations (due to reduction of molecular symmetry) or the presence of isomers and amorphic forms.

A helpful approach for description of vibrational spectra is to analyze them in terms of characteristic vibrations of functional groups. They occur in specific regions of a spectrum, e.g., the C-H stretching mode is observed in the region of $2850 - 3100$ cm^{-1}. Figure 1.2 shows typical vibrations for a functional group XY_2. Vibrations associated with the change of a bond length are called stretching vibrations, whereas those expressing change of the molecular angle are called deformational or bending vibrations. Another criterion arises from symmetry of vibrations (symmetric or asymmetric) or from motion in-plane and out-of-plane of a functional group (see Figure 1.2). Using tables of group frequencies, Raman and IR spectra can be described qualitatively. Each of these vibrations can be also considered in terms of principles of the group theory. Thus, the A-type vibrations are non-degenerate vibrations, symmetric to an axis of the highest multiplicity. The B-type vibrations are also non-degenerate vibrations, but anti-symmetric to an axis of the highest multiplicity. Finally, the E- and T(F)-type vibrations are vibration double and triple degenerated, respectively. Additional indexes placed with the symbol, such as 1, 2, 'i', 'g' (gerade) and 'u' (ungerade), determine the symmetry of vibrations with respect to an axis of multiplicity different than a major axis, a plane, the center of inversion, respectively.

Experimentally, the assignment of IR and Raman bands to vibrations of individual groups of atoms is also based on the isotopic shift (change of vibration frequency due to the substitution of a chosen atom by its isotope. This is supported by an assumption that the isotopic change does not alter distribution of electron density, and therefore it does not change the force constant of a molecule. Magnitude of this shift is determined by the equation 1.5:

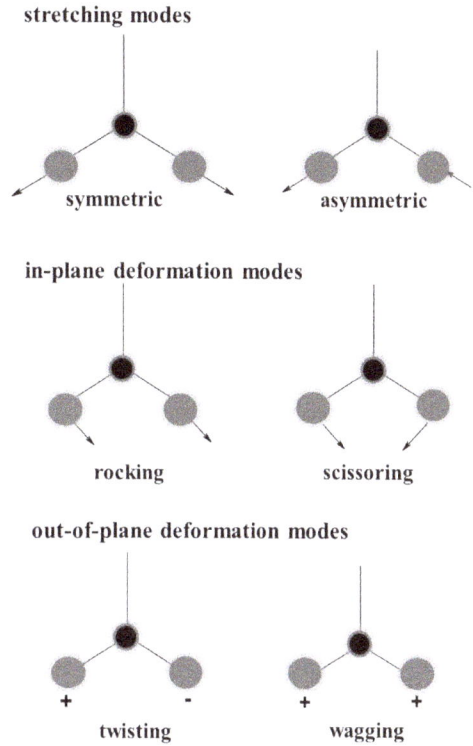

stretching modes

symmetric   asymmetric

in-plane deformation modes

rocking   scissoring

out-of-plane deformation modes

twisting   wagging

Fig. 1.2. Types of vibrations for XY_2 molecule

$$\frac{\bar{v}}{\bar{v}*} = \sqrt{\frac{\mu^*_{red}}{\mu_{red}}}, \tag{1.5}$$

where „*" corresponds to a molecule with a changed isotope.

It is also assumed that a considerable shift is only observed for bands that represent "isolated" vibrations of functional groups.

1.3. Construction of a FT-IR spectrometer

Fourier Transform IR spectrometers operate on the basis of mathematical operation called Fourier transformation. A typical spectrometer consists of a light source, an interferometer, a sample chamber and a detector.

1.3.1. Light source

A source of light in the IR spectrometer must be polychromatic infrared radiation due to the necessity of fitting its energy to transitions between vibrational levels of all the molecules. Typically, light sources are a globar (SiC) and a Nernst rod ($ZrO_2 + Y_2O_3$), which – when heated to high temperature – emit radiation similar to the black body radiation. The Nerst rod emits radiation in the range from 300 to 20 000 nm (from UV to NIR), whereas the globar provides light in the range of 1 100 – 40 000 nm (NIR, MIR and FIR). The light beam from the source reaches the interferometer, then is collimated through a series of mirrors and finally directed to a sample.

1.3.2. Interferometer

In the Fourier Transform spectrometers, intensity of absorption is not directly recorded as a function of frequency but in a form of an interferogram, which represents relationship between a signal and time (the optical path difference). To obtain a spectrum in the frequency domain from the time domain, the interferogram is transformed by a mathematical operation called the Fourier transformation. Collection time of a spectrum is very short due to the fact that the entire spectral range is recorded simultaneously. A number of spectra registered in the time domain are averaged, and then transformed by the Fourier transformation. Advantages of the use of the interferometer in any spectrometer are as follows:

1/ Felgett's adavantage (multiplex advantage): recording all wavelengths simultaneously results in shortening data collection time and improvement of signal (S) to noise ratio (N) since $S/N \sim n^{1/2}$ (n – number of spectra). This advantage is especially important for techniques which generate weak signal (e.g. a Raman spectrometer with laser excitation in NIR region),
2/ Jacquinot's advantage: the lack of slits does not limit radiation pathway,
3/ Connes's advantage: the movement of the moving mirror is controlled optically by a He-Ne laser and this allows to achieve high precision of the wavenumber scale in a spectrum.
4/ easy change of spectral resolution.

A schematic of a typical interferometer (the Michelson interferometer) used in optical spectroscopy is illustrated in Figure 1.3. The interferometer is an element splitting the electromagnetic radiation beam and its role is similar to a grating in the monochromator. It consists of two mirrors, positioned to each other at an angle of 90°, when one of them is stationary, whereas the second mirror moves controlled by a He-Ne laser. Between the two mirrors, a beam splitter is placed at an angle of 45°. The radiation falling on the beam splitter is divided into two beams transferred to both mirrors. Then light beams are reflected on the mirrors and next they are collimated on the beam splitter. Light beams interfere to each other depending on the position of the moving mirror. If the two mirrors are located at the same distance from the beam splitter, and hence the optical path difference is zero, the highest amplification of all wavelengths is observed. In case of any different positions of the moving mirror, both beams are no longer in-phase consistent and only those wavelengths for which the interference condition is satisfied will be amplified.

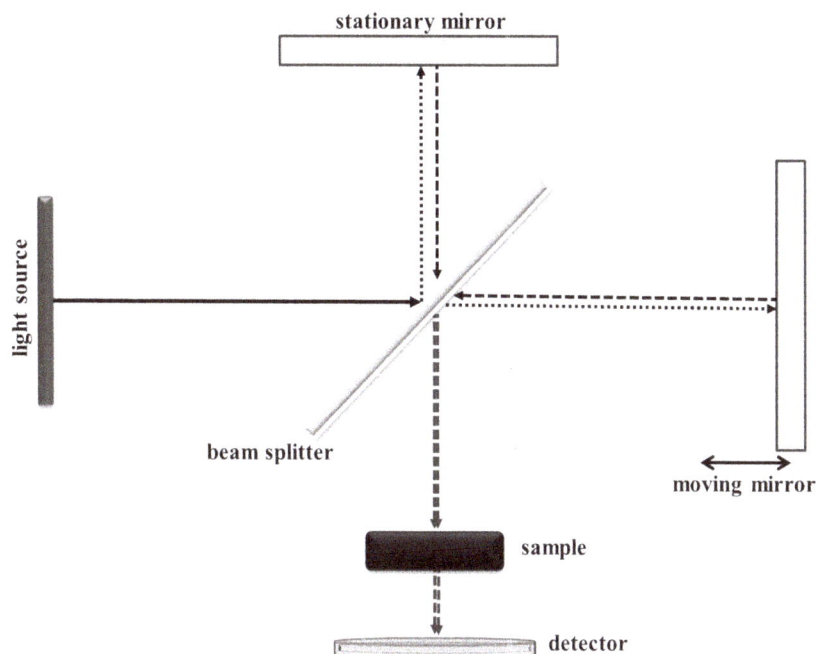

Fig. 1.3. A schematic of the Michelson interferometer along with marked pathway of beam rays

1.3.3. Sample chamber

A sample chamber of a FT-IR spectrometer is constructed specifically for each IR technique, discussed in Chapter 3.1. The chamber and the spectrometer are purged with dry air or dry nitrogen in order to remove water vapor and CO_2 present in air. Both molecules, due to the large dipole moment, generate strong bands in spectra. Water vapor shows bands at about 3500, 1630 and below 500 cm^{-1} whereas CO_2 absorbs infrared radiation at 2390 – 2280 cm^{-1} (doublet) and 672 – 665 cm^{-1}. Mostly, IR spectrometers are single beam instruments and thus collection of the empty chamber or a cuvette with solvent is required before measurement of the sample.

1.3.4. Detector

After passing through a sample, signal is analyzed by a detector, whose function is to change intensity of radiation into an electrical signal. The standard detector in FT-IR spectrometers is DLATGS (deuterated triglicyne sulfate substituted with alanine) operating in the range of 350 – 6000 cm^{-1}. This detector is a pyroelectric cell sensitive to temperature changes induced by IR radiation. In FT-IR microscope, a MCT (semiconductor – Mercury Cadmium Telluride) detector is commonly used which detects radiation in the range of 670 – 7 000 cm^{-1} (see Chapter 3.2).

References

1. Cotton F.A., *Chemical applications of group theory*, Wiley, 1990.

Fundamentals of Raman scattering spectroscopy

Kamilla Malek, Małgorzata Barańska, Kamila Kochan

Raman scattering spectroscopy is based on transitions between vibrational and rotational levels of molecules, which occur as a result of inelastic light scattering. This effect was described theoretically by A. Smékal in 1923 and confirmed experimentally by an Indian physicist Chandrashekhar V. Raman in 1928. For this discovery he was awarded with the Nobel Prize in physics in 1930. The same effect of light scattering is observed in the macro- and microscopic world.

One can imagine it like scattering of light by colloidal solution if the particle size is similar to the wavelength of the incident light. Light scattering is also observed on energy levels of molecules, and will be explained for the vibrational levels. Fundamental phenomenon of this spectroscopy is the excitation of molecular vibrations through illumination of a sample by light with frequency of v_0 from UV, Vis or NIR region. Quantum description of molecular vibrations is given in Chapter 1. A result of the "molecule – electromagnetic wave" interactions is scattered light with the same energy (Rayleigh scattering) and discrete components with frequency different than v_0 (Raman scattering). Thus, Raman spectrum consists of lines evenly located around a Rayleigh band with frequencies $v_0 \pm v_{osc}$ where v_{osc} corresponds to the transition frequency between vibrational levels. A magnitude of the Raman shift from the Rayleigh band <u>does not depend</u> on frequency of the incident radiation, but results solely from properties of scattering molecules. Raman bands appearing at the frequency lower than v_0 are called Stokes bands, whereas those at the higher frequency are anti-Stokes bands. Light scattering is a very weak effect and this is reflected by low intensity of bands. Intensities of Rayleigh and Raman scattering is approximately thousand and million times lower, respectively, than intensity of the exciting light. Most of Raman spectroscopy techniques employ measurements of the Stokes part of Raman spectra, presented in a scale of wavenumbers (expressed in Δcm^{-1}, so-called Raman shift) in the spectral region from 0 to 4000 cm^{-1}. Raman spectrum expressed in that scale corresponds to IR spectrum, what enables their comparison since both techniques are complementary to each other. Raman spectra, similarly to IR spectra, primarily provide encrypted information about molecular structure. They are used for identification and quantification of unknown substances and specific groups of atoms, studying of chemical reactions, as well as inter- and intramolecular bonding/interactions. Similarly to IR spectroscopy, Raman spectroscopy allows to investigate compounds in all physical states in a wide range of temperature and pressure.

2.1. Quantum description of Raman scattering

The classical theory proposed by Placzek explains the phenomenon of Raman scattering based on generation of scattered light by induced dipole moment (2.1), which is formed in molecules due to the action of incident electromagnetic radiation:

$$\mu_{ind} = \alpha E_0, \tag{2.1}$$

where α is a polarizability tensor, $\mathbf{E}_0$ – electric vector of electromagnetic wave. Polarizability represents the tendency of electrons of molecules to be distorted by an external electric field.

As we explained in Chapter 1, molecular vibrations are never "frozen" and from this reason polarizability tensor must be affected by normal mode. Hence, the induced dipole moment is also modulated by vibrations of molecules. Therefore, the dipole moment is given by (2.2):

$$\mu_{ind} = \alpha E_0 \cos 2\pi c \nu_0 t + {}^1\!/_2 \left(\frac{\delta\alpha}{\delta q}\right)_0 Q_{osc} E_0 \times [\cos 2\pi c(\nu_0-\nu_{osc})t + \cos 2\pi c(\nu_0+\nu_{osc})t] \tag{2.2}$$

This expression explains how the induced and also oscillating dipole moment simultaneously generates electromagnetic radiation of three frequencies, corresponding to Rayleigh (see Eq. 2.2), Stokes (in blue) and anti-Stokes (in green) Raman scattering. This theory also explains the independence of the Raman shift on the wavelength of excitation, and further determines a selection rule of Raman scattering phenomenon. Specifically, Raman spectrum exhibits the presence of vibrations for which at least one component of the polarizability tensor changes due to vibration.

The energy diagram in Figure 2.1 shows the two-photon nature of Rayleigh and Raman scattering resulting from quantum rules. Both effects can be described as two single-photon processes appearing simultaneously. Firstly, absorption of photon to a virtual state with energy lower than those for excited electron state occurs (for normal Raman scattering), and then emission (after $\sim 10^{-14}$ s) of another photon with the same (Rayleigh scattering), lower (Stokes Raman scattering) or higher energy (anti-Stokes Raman scattering) is observed.

Raman transitions are allowed when their transition moment is non-zero and is determined by the change of polarizability due to transition from n to m vibrational state (2.3):

$$\alpha_{nm} = \langle \Psi_n | \hat{\alpha} | \Psi_m \rangle \neq 0. \tag{2.3}$$

The quantum approach, as the only one, properly describes the intensity of the scattered radiation for a given $n \rightarrow m$ transition. The integral intensity of the Raman

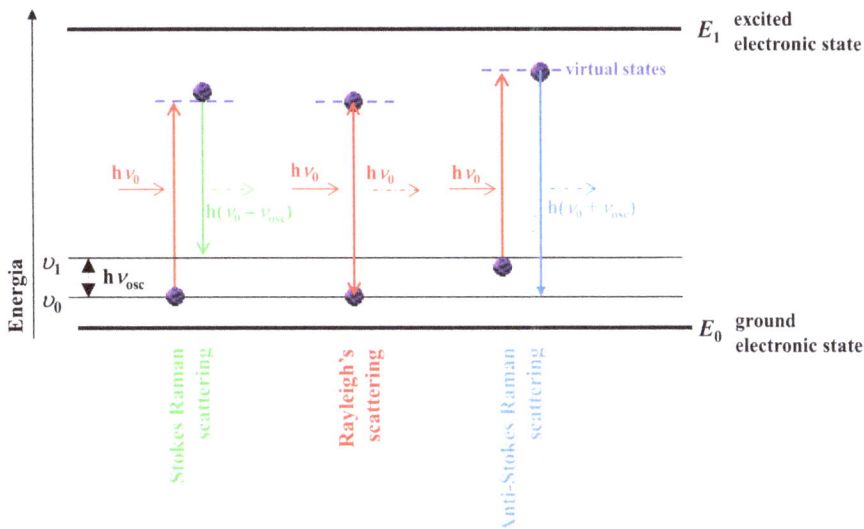

Fig. 2.1. A schematic of energy transitions for light scattering

radiation (I_{Raman}) depends on intensity of incident light (I_0), the fourth power of the frequency of the excitation light, and on the square of transition moment α_{nm} (2.4):

$$I_{Raman} \propto I_0 \cdot (v_0 \pm v_{osc})^4 \cdot |\alpha_{nm}|^2. \tag{2.4}$$

Population of molecules on a particular vibrational state and at a given temperature (Boltzmann distribution) determines Stokes and Anti-Stokes Raman intensity. At standard conditions, the ground vibrational level ($v=0$) is mainly occupied and thus the intensity of Stokes bands is higher than Anti-Stokes counterparts. Intensity ratio is expressed by (2.5):

$$\frac{I_{Anty-Stokes}}{I_{Stokes}} = \left(\frac{v_0+v_{osc}}{v_0-v_{osc}}\right)^4 exp\left(-\frac{hv_{osc}}{kT}\right), k - \text{Boltzmann constant}, T - \text{temperature} \tag{2.5}$$

For UV and Vis excitation $v_0 >> v_{osc}$, so the part of the equation 2.5 is equal unity.

Each molecule possesses the non-zero polarizability, resulting from structural features of a molecule. Polarizability is a 3×3, rank two tensor (represented by a 9-component matrix; Eq. 2.6) and if all the components of tensor are identical, a molecule is isotropic. In case of molecular anisotropy, at least one tensor value is different from the others (e.g. $\alpha_{XX} \neq \alpha_{YY} = \alpha_{ZZ}$).

$$\alpha = \begin{pmatrix} \alpha_{XX} & \alpha_{XY} & \alpha_{XZ} \\ \alpha_{YX} & \alpha_{YY} & \alpha_{YZ} \\ \alpha_{ZX} & \alpha_{ZY} & \alpha_{ZZ} \end{pmatrix}. \tag{2.6}$$

Intensity of Raman active vibrations is dependent on incident light polarization and this is expressed by the depolarization ratio ρ (2.7). The depolarization ratio is determined by recording the intensity of scattered light polarized perpendicularly ($I_{90°}$) and the intensity of scattered light polarized parallelly ($I_{0°}$).

$$\rho = \frac{3\beta^2}{45\alpha^2 + 4\beta^2} = \frac{I_{90°}}{I_{0°}}, \beta - \text{anisotropy of the polarizability tensor} \qquad (2.7)$$

The depolarization ratio allows determining the symmetry of normal mode (vibration) and its value varies from 0 to 0.75. For $\rho = 0$, a vibration is totally symmetric (e.g. A_1) and its band is very intensive and called a totally polarized band. In turn, when ρ is 0,75 a vibration is totally asymmetric or degenerated (e.g. B_2, T). The corresponding bands are generally of low intensity and are called depolarized bands.

2.2. Instrumentation in Raman spectroscopy

Raman spectrometers are divided into dispersive and interferometric instruments. Examples of commercially available spectrometers are shown in Figure 2.2 and their essential components are compared in Table 2.1.

Table 2.1 clearly shows that the fundamental difference between the dispersive and interferometric spectrometers regards an optical element/setup, whose function is to split light scattered by a sample. Individual components of a Raman spectrometer are described below.

Table 2.1. A comparison of dispersive and interferometric Raman spectrometers

Dispersive spectrometer	Interferometric spectrometer
Incident laser	
UV, Vis, NIR (wavelength up to 830 nm)	NIR (usually the 1064 nm wavelength)
Sample compartment: microscope stage; *x, y, z* – stage; macro chamber	
Optical element splitting light	
Monochromator (diffraction grating)	Interferometer
Detector	
CCD camera	Semiconductor detector

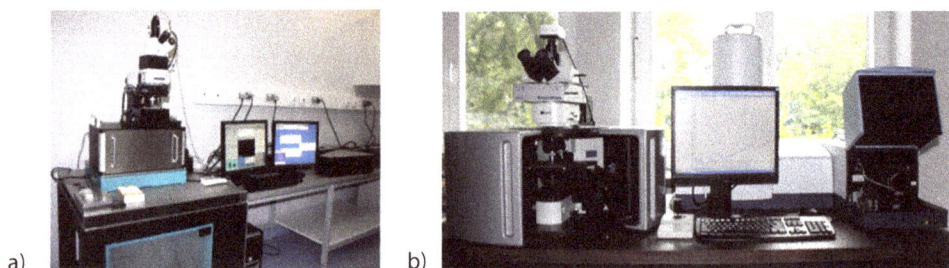

a)   b)

Fig. 2.2. A. Dispersive Raman spectrometer (WITec); B. Interferometric Raman spectrometer (Bruker)

2.2.1. Laser

The introduction in the 60s of the twentieth century laser (light amplification by stimulated emission of radiation) as a light source in the Raman spectrometer contributed undoubtedly to the spread of this technique due to the fact that such source is characterized by high intensity. The type of a laser is usually determined by the type of the gain medium, which generates a given wavelength of light. This may be a gas (e.g. an argon laser, Ar^+) or their mixture (e.g. a helium-neon laser, He-Ne), a liquid (dye lasers) and a solid (e.g. Nd:YAG and ruby lasers). The main part of the laser is a resonator – a system of two mirrors. One of them is impermeable (reflectance, $R \approx 100\%$) and the second mirror is partially transmissive ($R \approx 95\%$), which allows for output of a laser beam. Such a constructed optical resonator generates a monochromatic, coherent, parallel and polarized light beam of variable power and continuous or pulsed working mode. If the laser is able to emit a few exciting lines, it is called tuneable. Dispersive Raman spectrometers are usually equipped in lasers emitting ultraviolet, visible or near infrared light. However, NIR lasers with an excitation less than 830 nm are rarely used because of low quantum efficiency of a CCD detector in this range. The most commonly used lasers are those working in a Vis region, e.g. an argon ion laser (Ar^+) with excitation wavelengths at 488 and 514.5 nm, a helium-neon laser (He-Ne) with the 632.8 nm excitation line and diode lasers with an excitation at 785 and 830 nm (a NIR region). In turn, the interferometric spectrometer usually employs a Nd:YAG laser (yttrium aluminium garnet doped with neodymium Nd^{3+} ions; $Y_3Al_5O_{12}+Nd_2O_3$) operating in the continuous mode in near infrared, $\lambda = 1064$ nm. The main advantage of using a NIR laser comparing to lasers emitting UV-Vis light is eliminating fluorescence background in Raman spectra since most organic molecules exhibit fluorescence features in the UV-Vis range. However, as shown in Eq. 2.4, Raman intensity decreases along with the increase of wavelength of the incident light. This loss is compensated by the use of the interferometer (*see* Chapter 1).

2.2.2. Sample compartment

Exemplary sample compartments are illustrated in Fig. 2.2. Collection of Raman spectra does not require a complicated sample preparation procedures or a special type of sample cuvettes. Most Raman spectrometers are equipped with a microscope stage on which a sample is directly placed and measurements are performed *in situ* (Fig. 2.2. A). The fiber optics can output laser light out of the spectrometer and then record Raman spectrum from a distance of several meters while a sample container is insensitive to laser light, spectra collection of a toxic or dangerous substance is carried out directly in a container. For special measurements (*e.g.* investigations of chemical reactions), a special flow cell is used, in which reactants can be easily mixed. The important advantage of Raman spectroscopy is the possibility to record vibrational spectrum of aqueous solutions because polarizability of water is low in contrary to its dipole moment. From this reason IR spectrum is dominated by water bands. The main problem which can obscure Raman signal is the presence of fluorescence background since this phenomenon often accompanies Raman scattering effect. A removal or reduction of the fluorescence background is realised in several ways such as addition of fluorescence quenchers (KBr, NaI, KI); the use of pulse excitation; recording the Anti-Stokes part of Raman spectrum; a change of an excitation wavelength and laser irradiation of a sample before collection of Raman spectrum.

2.2.3. Optical element splitting the beam of electromagnetic radiation

A monochromator is an optical system used to split a narrow region of light wavelengths and is used in the dispersive Raman spectrometers. The monochromator construction is identical in the entire region of light but differs in materials used for fabrication of optical elements (slits, lenses, mirrors, windows, prisms or diffraction gratings). Construction of a monochromator is as follow:

- an entrance slit through which light beam enters perpendicularly,
- a set of lenses or mirrors, which produce a parallel beam of light,
- a diffraction grating (or a prism), which disperses radiation on its components,
- a collecting element, which converts the image on the input slit into an image on the focal plane,
- an exit slit, which transfers a chosen range of electromagnetic radiation to a detector.

An essential element of the monochromator is the diffraction grating. It is regularly spaced set of parallel slits. The slits cause deflection of the beam and its diffraction occurs. An example of light diffraction on the grating is a CD disc. A typical number of slits per 1 mm used in Raman spectrometers varies from 300 to 2000 for UV-Vis light and 10 – 200 for IR. To change spectral resolution in the dispersive Raman spectrometer a grating must be replaced and the instrument must be recalibrated.

The use of a Nd:YAG laser, operating in the near infrared region, allows to "escape" from fluorescence but it significantly reduces Raman intensity by a factor of v_0^4. From this reason, Raman spectrometers are equipped with an interferometer instead of a monochromator since the interferometer improves signal to noise ratio (S/N). The principles of the action of an interferometer are described in Chapter 1.3.2.

2.2.4. Detector

Dispersive and interferometric Raman spectrometers require to use different detectors because their quantum yield and sensitivity should be matched to wavelength of Raman scattered light. A CCD camera (Charge-Coupled Device), a multi-channel photo-emitting detector, is usually employed in the dispersive Raman spectrometers. It is a two-dimensional diode line that "reads" each component (intensity and wavelength) of the spectrally splitted light by the monochromator. A CCD camera exhibits a high quantum yield in the range of 300 – 900 nm. For example, a 512×320 CCD camera consists of 163 840 detectors (pixels) arranged on the surface of 6.5×8.7 mm^2. In interferometric Raman spectrometers, on the other hand, semiconductor detectors are usually used as they show a high sensitivity in the near infrared region. The most commonly NIR detectors are the InGaAs-type detector working at room temperature and a very sensitive germanium detector working at the liquid nitrogen environment.

3

Special FT-IR techniques

3.1. Sampling techniques in FT-IR spectroscopy

Paweł Miśkowiec

The history of infrared spectroscopy reaches 1882, when Sir William Abney and Edward Festing, photographed for the first time absorption spectra in the near infrared range of over fifty organic liquids and correlated the absorption bands with the presence of specific functional groups in the molecule. However, it is William W. Coblentz, who is considered to be the father of infrared spectroscopy. He had built on his own the first prototype of the IR spectroscope in 1903 and until 1905 he registered the spectra of over a hundred organic and inorganic compounds also in the mid-infrared. The rapid development of spectroscopic methods in the infrared range had been achieved by implementation of Fourier transform (FT-IR) in the late forties of the 20th century as well as the development of computational techniques in the sixties of the 20th century which shortened time of the FT-IR measurements to the order of seconds. The application of FT-IR technique allowed to obtain several times higher signal to noise ratio compared to a conventional dispersion method.

The ability of weak signal registration from the sixties of 20th century has broadened significantly the range of available IR techniques of spectra registration. Among currently used IR methods, besides the transmission technique, one should mention: attenuated total reflection (ATR), Diffuse-Reflectance (DRIFTS), reflection-absorption infrared spectroscopy (IRRAS), photoacoustic spectroscopy and the emission spectroscopy (ES).

3.1.1. Transmission technique

The transmission technique is historically the oldest method of measurement in the infrared range. The polychromatic IR radiation of intensity I_0 passes through the sample and is absorbed. The detector records the intensity of radiation reduced by the part absorbed by the sample. Parameters used to determine the absorption of radiation in this technique are interchangeably absorbance (A) and transmittance (% T). The relationship between them is expressed in a well-known equation:

$$A = \log_{10}(100/\%T), \qquad (3.1.1)$$

Transmittance is expressed by the relation:

$$\%T = I/I_0. \qquad (3.1.2)$$

Absorption of radiation occurs in accordance with the Beer-Lambert law, that is why the transmission technique is used successfully both for the qualitative and quantitative determination.

In the transmission technique the sample can be analysed in the gaseous state (in the gas cuvettes), liquid, or solid states. In the case of a solid, sample usually has to be previously prepared either in the form of a thin film or crushed and mixed with material permeable for IR to form a crushed powder or a pellet.

Two dilution compounds are usually used in preparing the suspensions: paraffin oil (nujol) and Fluorolube® (chlorotrifluoroethylene). The typical methods of preparing a suspension of the sample is based on grinding with a small amount of oil until formation a paste, which subsequently forms a thin film between the two windows mostly made of KBr. Alternatively, the analysed substance can be dissolved in one of typical solvents used for IR spectroscopy and studied in special cuvette dedicated for IR spectroscopy. One of the best solvents (most transparent for infrared radiation), are halogenated derivatives of hydrocarbons such as CCl_4 or C_2Cl_4. Apart from them, one frequently uses the above mentioned nujol, as well as carbon disulphide, chloroform, dichloromethane or acetonitrile.

The cuvettes used in the infrared spectroscopy, as well as spectrometers' optical elements must be made of material permeable to infrared radiation. Unfortunately glass is an impermeable material for radiation with wavelength greater than 2.5 microns (less than 4000 cm^{-1}), therefore cannot be used in the mid-infrared spectroscopy. The most common and cheap materials used in IR spectroscopy are NaCl and KBr. A disadvantage of these materials is their hygroscopicity and solubility in water. Therefore they cannot be used in the analyses of aqueous solutions and have to be protected from atmospheric moisture. Aqueous solutions can be studied using cuvettes made of water-insoluble materials such as SiO_2, CaF_2, BaF_2, AgCl, or KRS-5*.

3.1.2. Reflexive techniques

Nowadays the most frequently used reflection techniques are described with the following acronyms: ATR, DRIFT and IRRAS.

---

* KRS-5 – thallium bromoiodide crystal

In the ATR technique, the well-known optical phenomenon called the total internal reflection is employed. The radiation beam from the material of refractive index n_1 incidents the material of refractive index n_2 at an angle α. While penetrating the second material the beam is diffracted by the angle β. When $n_1 > n_2$, then $\beta > \alpha$. According to the law of refraction, the total internal reflection occurs when $\beta = 90°$. Then the value of the critical angle of incidence of the beam is depicted with the equation:

$$\alpha_c = \arcsin\left(\frac{n_2}{n_1}\right). \qquad (3.1.3)$$

Thus, when $\alpha > \alpha_c$, the light undergoes total internal reflection from the border between both materials at the angle α.

As a result of the interference between the waves incident and reflected the standing wave is forming and propagating in the direction perpendicular to the border between both materials. Thus, the standing wave penetrates also the substance with a lower refractive index. If this material absorbs infrared radiation, the intensity of the standing wave is weakened because of the absorption. In practice, in the ATR method one use the crystal transparent to infrared radiation in the maximum range and a high refractive index n_{ATR}. In order to properly measure the ATR signal, sample must fit tightly to the crystal. For samples with an uneven surface the special mechanisms are used to push them thoroughly to the crystal. The amplitude of the standing wave decreases exponentially with the distance. The penetration depth of the sample (d_p) with the standing wave depends on the refractive indices of the crystal (n_{ATR}), of the sample (n_{sample}) and the radiation incidence angle α for a given wavelength λ. This relationship is expressed by the formula (3.1.4):

$$d_p = \frac{\lambda}{2\pi n_{atr}\sqrt{\sin^2\alpha - (\frac{n_{sample}}{n_{ATR}})^2}}. \qquad (3.1.4)$$

Typically, in the ATR measurements d_p does not exceed 2 μm. The technique is primarily a spectroscopic method for testing of the surface.

As the penetration depth depends on the wavelength of the incident beam, the recorded intensity of the beam is a function of wavelength too. If the ratio n_{sample}/n_{ATR} is fairly stable in the whole range of observed wavenumbers, an increase in the wavelength of IR radiation increases the penetration depth. This phenomenon requires an amendment on the relative intensities of the bands. The difference in the profile of the spectrum before and after the ATR amendment is illustrated in Fig. 3.1.1.

With a single reflection ATR method's sensitivity is relatively low. That is why the multiple reflections are widely used. The intensity of absorption increases proportionally to the number reflections of the beam in the crystal. The principle of the method both in single and multi-reflection variants is shown in Fig. 3.1.2.

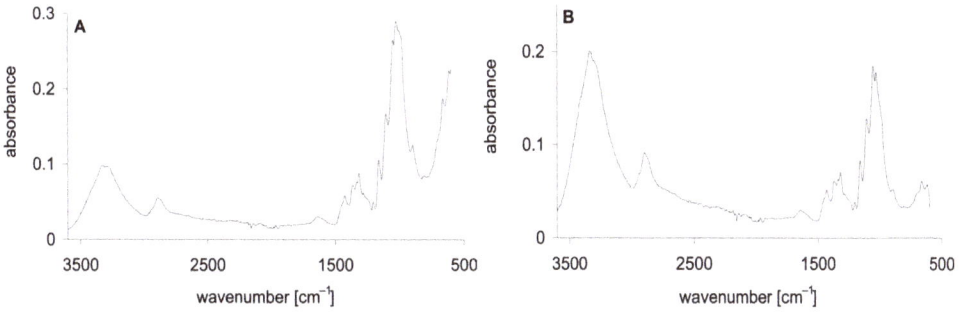

Fig. 3.1.1. ATR FT-IR spectrum of cellulose before (A) and after the application of amendment on the relative intensities of the bands (B)

Spectral range of the ATR technique depends on the permeability of the crystal. Typical ATR crystals and their characteristics are collected in Table 3.1.1.

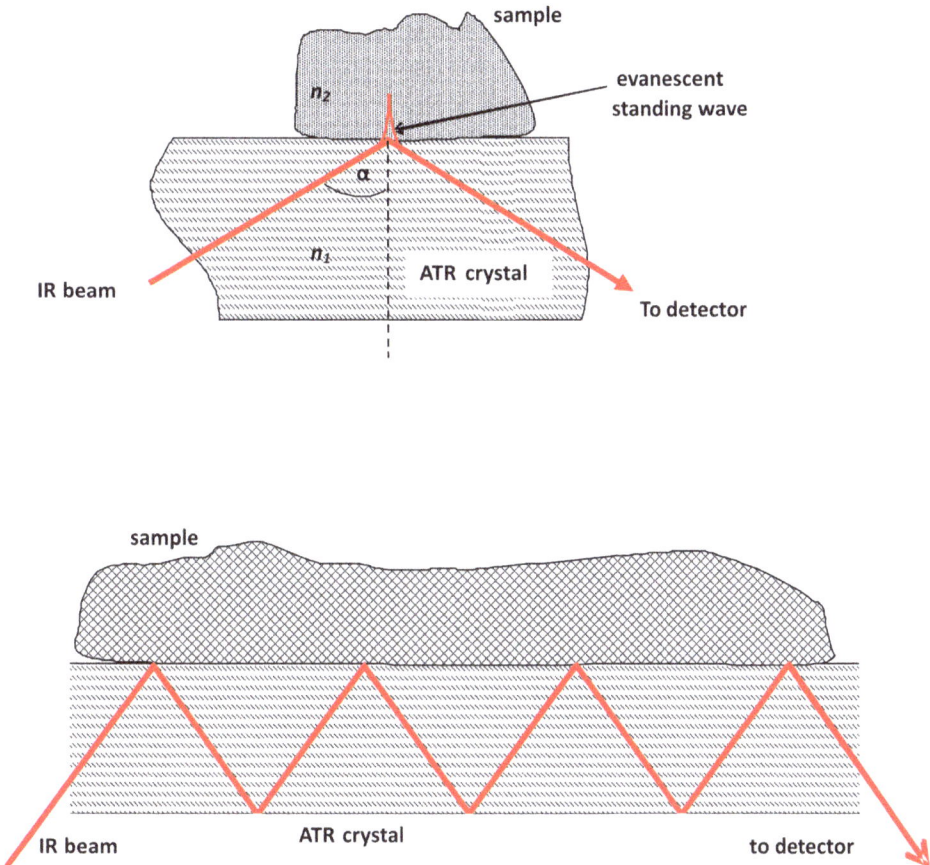

Fig. 3.1.2. The scheme of penetration of the sample with the IR beam in the single and multi--reflection variants

Table 3.1.1. The properties of crystals commonly used in the ATR technique

Material	Spectral range [cm^{-1}]	Refractive index	Application
Germanium	5 500 – 400	4.00	Durable, for all kinds of materials
Silicon	8 900 – 660	3.41	Resistant to the most of the solutions at pH 1–12
ZnSe	17 000 – 650	2.40	Mostly for soft materials and solutions of pH 5–9
Diamond	30 000 – 2 200 2 000 – 400	2.41	Suitable in particular for hard materials
KRS-5	20 000 – 250	2.37	Soft, toxic, less frequently used
AMTIR-1*	11 000 – 700	2.50	Resistant to acids

* AMTIR-1 – Amorphous Material Transmitting Infrared Radiation. It is a glass having a composition $Ge_{33}As_{12}Se_{55}$.

In the ATR technique the physical morphology of the studied materials is usually not a difficulty, as long as there is a close contact provided between the surface of the sample and the crystal. With the ATR method one can measure irregular surfaces (e.g. powders and textiles), which in this technique give spectra of a good quality in a wide spectral range. Moreover, it is possible to register spectra *in situ* and even *in vivo*, what is extremely important for biological and medical research. The disadvantage of this method is a problem of reproducing of the contact surface between the sample and the crystal what causes a number of difficulties in quantitative analysis. Depending on the applied crystal one has to take into account crystal's susceptibility to mechanical and chemical damage. However, due to the fact that some samples may change their spectral properties as a result of preparation (grinding, pressing, etc.), ATR spectroscopy method can be an excellent alternative for the transmission technique, as the sample requires minimum preparation before the measurement.

3.1.2.2. Diffuse reflection in the mid-infrared range, DRIFT

Another of the reflection techniques is the method of diffuse reflection in the range of mid-infrared. It is used most often for measurement of spectra of solids or powders, for which the use of classical transmission technique is difficult or even impossible, due to poor absorption, strong scattering effect or limits in preparing of KBr pellets.

In this technique, the infrared radiation incidents on the sample and is reflected in the two ways: specular reflection in which the angle of incidence equals the angle of reflection, and diffuse reflection, wherein the light beam is reflected in different directions and at different angles than the angle of incidence. The dispersed component of the reflected light beam is formed due to the penetration of infrared radiation into the sample. The radiation is then absorbed and the light beam is deflected several times. While reaching again the sample surface such a beam is radiated in all directions (Fig. 3.1.3).

specular reflection   incident beam

diffuse reflection

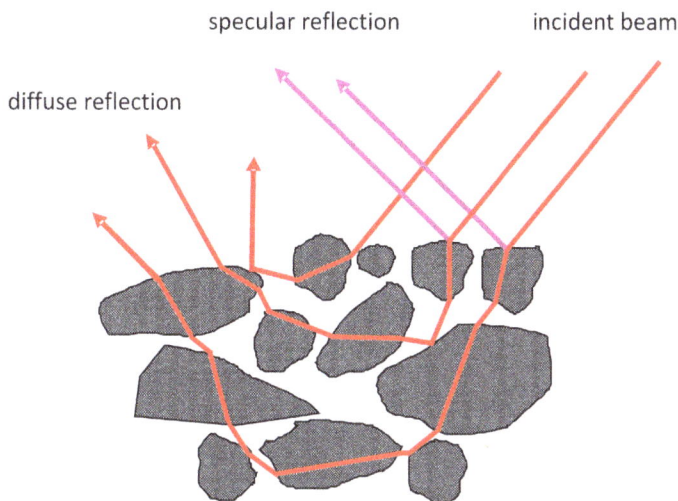

Fig. 3.1.3. The mechanism of diffuse radiation and the reflected mirror

The radiated light is collected with one or several mirrors and directed to the detector (Fig. 3.1.4). In addition, the shutters are used, which eliminate the specular reflection. The obtained spectrum is a dependence of the reflectance on the wavelength. If the sample is sufficiently shredded or grated, specular component is infinitely small and can be neglected. Then the reflectance spectrum is characterized only by scattered radiation and is therefore comparable to the transmission spectrum. The theoretical basis of the phenomenon of diffuse reflection is described by

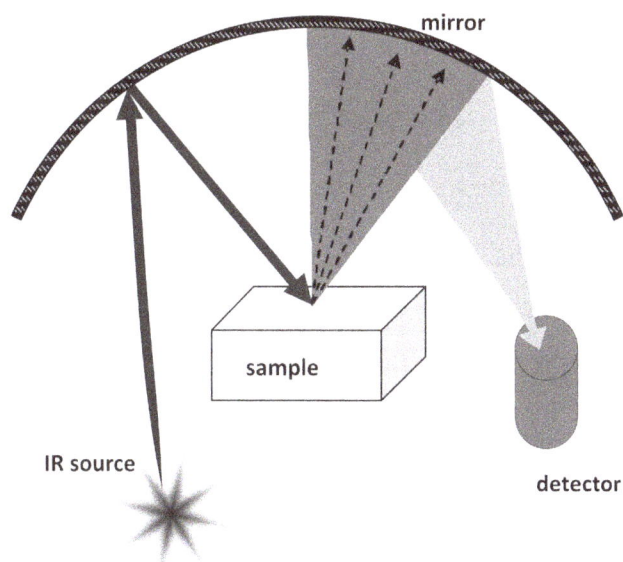

mirror

IR source

sample

detector

Fig. 3.1.4. Scheme of DRIFT accessory

the theory of Kubelka-Munk. According to this theory, in the case of an infinitely thick layer of a substance, which slightly absorbs infrared radiation, an amount of the scattered radiation is a function of the concentration of the studied substance. Kubelka-Munk (KM) function has the form:

$$KM = (1 - R_\infty)^2/2R_\infty = K/S, \tag{3.1.5}$$

where R_∞ – reflectance factor of the layer, K – absorption coefficient, S – scattering coefficient.

Thus, Kubelka-Munk function is given as a ratio of the two factors: K and S. In case of low concentration of analyte, K-factor can be written as: $K = 2.303ac$, where a – the absorption capacity of the analysed substance, c – concentration of the absorbing compound. The equation (3.1.5) takes the following form:

$$KM = 2.303ac/S. \tag{3.1.6}$$

This equation is the basis of quantitative analysis based on the DRIFT technique.

Please note that the above linear relationship is fulfilled only if:

- Analytes are much diluted in the matrix non absorbing for IR radiation (e.g. KBr, KCl, etc.)
- Sample is homogenously and firmly grinded (grain size of <2 μm), because S-factor strongly depends on grains size and the level of grains compression.
- Thickness of the layer of the sample equals at least 1.5 mm,
- The value of specular reflection is negligibly small.

Taking into account the above mentioned rules, one can employ the DRIFT method for quantitative analysis up to concentrations of ppm.

3.1.2.3. Reflection – absorption infrared spectroscopy, IRRAS

Reflection – absorption infrared spectroscopy is a technique used to study membranes and monolayers relatively weakly absorbing infrared radiation and deposited on a carrier which have a high reflectance factor of infrared ray.

When the radiation beam reaches the surface of the highly reflective sample at the wide angle, a standing wave is generated. The electric field vector of this wave is perpendicular to the reflecting surface. Standing wave appears as result of interference of the parallel components of the incident and reflected radiation beams (Figure 3.1.5).

In the case of the application of a film or monolayer onto a carrier, in addition to the absorption of incident and reflected wave the generated standing wave is absorbed too.

The optimum angle of radiation depends on the type of reflecting surface and the type of the sample. It generally has a value between 75° and 85°. The perpendicular component of the incident beam, as not contributing information, is removed

perpendicular-polarized radiation components

$E_\perp$ $E_\perp$

reflecting surface

parallel-polarized radiation components

E

$E_\parallel$ $E_\parallel$

reflecting surface

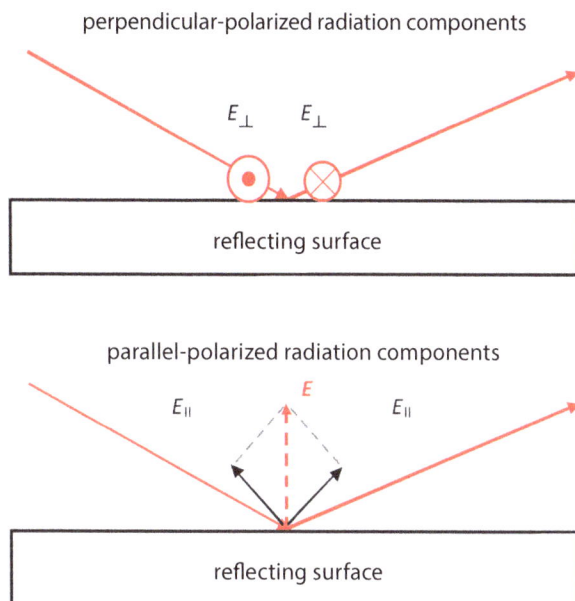

Fig. 3.1.5. The scheme of formation of standing wave. For radiation beam components perpendicularly polarized the intensity vectors of the incident and the reflected beams practically cancel out. In the case of the parallel polarized components the intensity vectors sum up to give a standing wave of the intensity E and perpendicular to the reflecting surface

by using a polarizer. Low intensity of the analysed signal forces the application of highly sensitive detectors, *i.e.* MCT.

An important feature of the IRRAS technique is the possibility to obtain information about the orientation of the particles in the analysed layer. Only vibrations of those functional groups will be excited which dipole moment of transition has a component perpendicular to the reflecting surface (Figure 3.1.6).

Spectra obtained with the IRRAS technique are comparable with the spectra obtained with transmission technique if the thickness of the sample is comparable to the wavelength of the incident radiation. In other cases, the spectrum should undergo a mathematical transformation. The most commonly used is the Kramers-
-Kronig transformation.

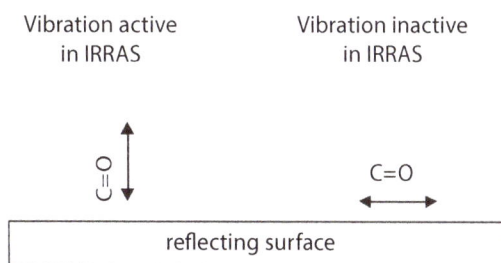

Vibration active
in IRRAS

Vibration inactive
in IRRAS

$C=O$

$C=O$

reflecting surface

Fig. 3.1.6. The effect of orientation of the functional groups on the activity of the vibrations in the IRRAS technique

Noteworthy variation of the IRRAS technique is a method with modulated polarization PM-IRRAS (*Polarization Modulation Infrared Reflection Adsorption Spectroscopy)*, which allows for practically complete elimination of a background signal originating from water vapour or carbon dioxide. Such devices may be disposed for instance, above the Langmuir bath, which allows studying monolayers and record *in situ* spectra.

3.1.3. Photoacoustic spectroscopy, PAS

One of the more unusual spectroscopic techniques is the photoacoustic spectroscopy. It is based on the photoacoustic phenomenon discovered in the eighties of the 19th century. In this method, the sample is irradiated with a modulated beam obtained, in the case of infrared range, from the incandescent lamps, Nernst lamp or a laser. As a result of irradiation particles transit to the excited state whereupon return to the ground state losing energy in a number of ways. One of these is nonradiative transition which causes a local change in temperature and as a consequence increases of the pressure of the gas surrounding analysed sample. Modulation of irradiation intensity causes pressure fluctuations, that is, the formation of the acoustic wave, recorded by special measuring microphones.

The main part of the photoacoustic spectrometer is a measuring chamber, with a sample of analysed material and the microphone inside (Figure 3.1.7.). The characteristic of the system generates the atypical needs and working conditions. First of all, the measuring chamber must be acoustically insulated from external audio

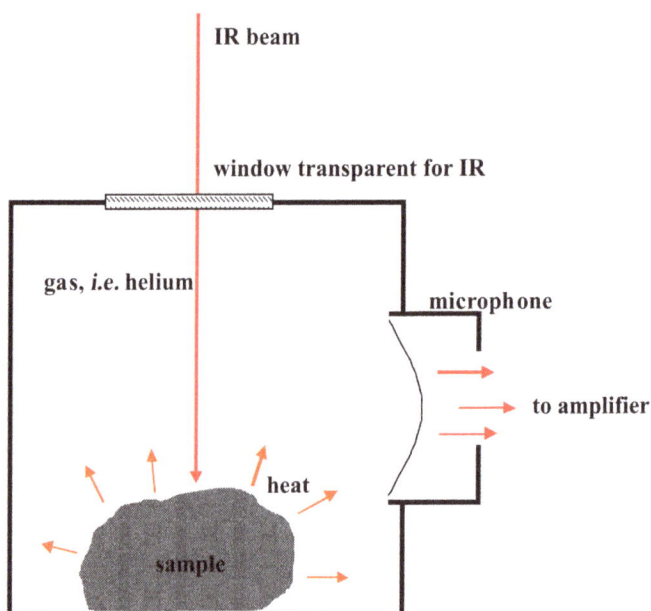

IR beam

window transparent for IR

gas, *i.e.* helium

microphone

to amplifier

heat

sample

Fig. 3.1.7. Diagram of the PAS chamber

sources. Its construction must ensure minimizing any internal reflections of radiation. Even the method of fastening of the sample affects the signal quality.

The PAS method was in the first place applied to the analysis of gaseous samples. However, it also allows obtaining IR spectra similar to the transmission and reflection for the substance of any states of matter and forms as powders, gels, colloids, including those with a very high absorption factor, for which conventional spectroscopic methods fail. This method, as a non-destructive one, is also used in medicine and biology, to study samples *in situ*.

3.1.4. Infrared emission spectroscopy, ES (IRES)

An alternative method for transmission and reflection spectroscopy is the infrared emission spectroscopy. Theoretical background of the above mention method is based on the Kirchhoff's law of thermal radiation, according to which the emissivity of the material body is proportional to the absorption capacity of the same body. Consequently, the emissivity is dependent on both temperature and frequency of the emitted radiation. However, IR emission spectroscopy is not widely used because the emission signals from the sample are relatively weak, overlapped with the ambient radiation and propagated isotropically.

In the ES method, the sample must be heated to a temperature higher than the temperature of the detector. Usually while sample heating, the detector is being cooled to avoid its own emission. The heated sample emits infrared radiation, the intensity of which depends on the number of molecules on the excited vibrational levels according to the Boltzmann distribution. Quantitative analysis in the emission method is in principle similar to that of the transmission technique. However, self-absorbing of the emitted radiation by the sample should be taken into account.

IR emission spectroscopy has been successfully used in the analysis of substances which cannot be illuminated, as well as in the monitoring of production processes. The sample in this method does not require any special preparation, a separate source of infrared radiation is not necessary either. This method may be particularly useful in the analysis of low frequency vibrations (<1000 cm^{-1}), for which the absorption and reflection spectra are complicated to interpret. Note that the results strongly depend on the type of sample, its thickness, geometry and refractive index. The aforementioned heat radiation of all the surrounding parts warmer than the detector affects also the quality of the spectra. Therefore, the thermal shielding is crucial as well as taking into account the impact of the ambient. On the other hand, the emission spectroscopy can be successfully used to analyse the impact of factors discussed above on the sample (*i.e.* in the studies of catalytic processes or in the electronic industry).

3.2. FT-IR microscopy and imaging

Ewelina Wiercigroch, Kamilla Malek

The coupling of a FT-IR spectrometer with a Cassegrain objective microscope has introduced a powerful tool for simultaneous analysis of the chemical structure and spatial distribution in heterogeneous materials of micron size (~2.5–25 μm). An infrared microscope is an optical system designed for two purposes: to allow the user to see small (micron-sized) samples and to obtain accurate infrared spectra on those small samples. Infrared microscopy can be done in three modes: 1/ single point, 2/ mapping by collecting spectra from single point or linear array and 3/ 2-D Focal Plane Array (FPA) imaging. They are showed in Fig. 3.2.1. The type of FT-IR microscopic technique depends on a detector. The aperture setting in a IR microscope determines the sample's region of interest in single mapping technique. Here, single point MCT detectors, which operate in IR microscopes, usually have a size of 50×50, 100×100, or 250×250 μm². Then, a spectrum is recorded and a x,y,z stage moves to the next region. The collected spectrum represents average chemical information from the selected area.

The spatial resolution of the chemical image is inherently affected by the diffraction limit according to Rayleigh's criterion: $0.61\alpha/NA$. The latter depends on the numerical aperture (NA) of the objective and the light wavelength (λ). The largest numerical aperture for commercially available Cassegrain objectives used in IR

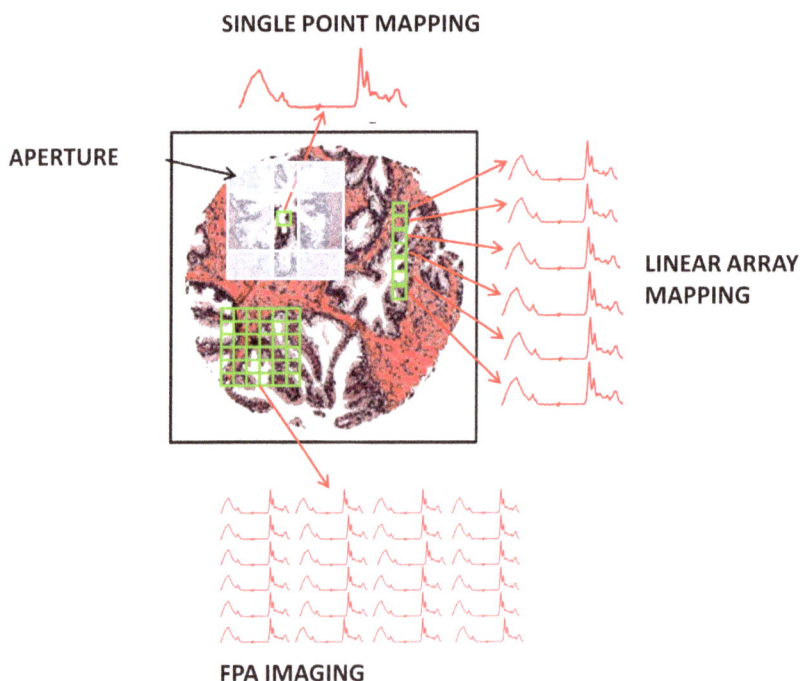

Fig. 3.2.1. A scheme illustrating single point, linear array mapping and FPA FT-IR imaging

microscopes is *ca.* 0.65. Since a light source employed in IR microscopes is poly-chromatic, the spatial resolution of IR images is determined by the wavelength. Hence for the typical spectral range from 4000 to 700 cm^{-1}, the diffraction-limited spatial resolution varies from 1.7 to 14.3 μm. In turn, a linear MCT detector consists of a 16×1 MCT array (400 μm × 25 μm). Each element of the linear detector records the spectrum of the corresponding projected pixel size onto the sample. A spatial resolution for this mode is similarly to single mapping limited by the Rayleigh's criterion. Only a matrix of MCT detectors (MCT FPA, MCT Focal Plane Array) enables collection of hyperspectral imaging. A MCT FPA detector is built of up to 128×128 elements, which give 5.5×5.5 μm^2 of the pixel projection onto the sample plane. This value is a limit of the achievable spatial resolution in FTIR imaging. The 128×128 MCT FPA detector allows simultaneous collection of 16,384 spectra from all positions within the sampling area of *ca.* 700×700 mm^2 in one tile (one "snapshot"). The FPA system is apertureless, so diffraction effects are limited, contrary to mapping microscopes.

IR microscopes are designed to operate in either transmission or reflection mode. In addition, the microscope can be equipped with a micro-ATR accessory (usually germanium crystal) allowing collection of attenuated total reflection IR spectra (ATR FT-IR). Fundamentals of FT-IR techniques are discussed in detail in Section 3.1. A schematic of the available techniques is depicted in Fig. 3.2.2.

The best S/N is found for the transmission mode, this technique is the most common for IR microscopy, details in Section 3.1. However, the sample must be placed on a substrate made of a material permeable for IR light such as non-hygroscopic CaF$_2$, BaF$_2$ or CsI windows. To fulfil the Lambert-Beer relationship, the thickness of a sample should be adjusted, commonly it does not exceeds *ca.* 10 μm. The IR windows are relatively expensive material and from this reason an alternative

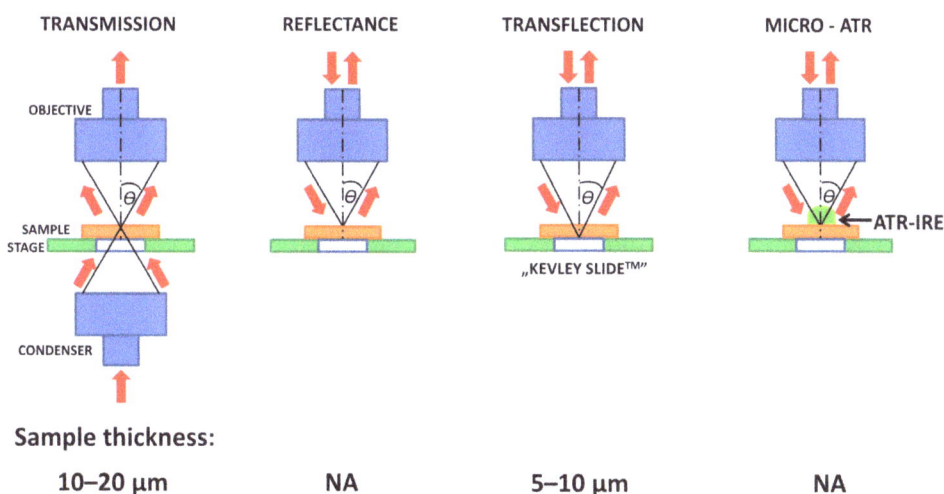

Fig. 3.2.2. The comparison of measurement modes used in FT-IR microscopy; red arrows shows IR radiation. Illustration of courtesy of Dr. M. Kansiz and Agilent Technologies

is the transflection mode. Here, Ag/SnO$_2$ coated infrared reflective slides (Kevley Technologies) are often employed. The beam of IR light passes through the sample and reflects off the IR reflective layer. The beam thus passes through the section twice, and consequently a suitable thickness of the samples is usually less than for the transmission technique. From one hand, this operation mode improves absorbance signal for very thin samples and along with the low cost of the reflective substrates in the comparison to the permeable IR widows make the transflection technique a useful tool for IR microspectroscopy. However, FT-IR spectra, especially biological materials, can be strongly distorted by the electric field standing wave (EFSW) effect. This results in the non-linear relationship between absorbance and the sample thickness causing significant changes in shapes and ratios of IR bands.

Similarly to typical FT-IR techniques, IR microscope can be used to record IR signal reflected from the surface of the sample (see reflectance mode in Fig. 3.2.2). In this case, the sample is placed on a highly-reflective material, e.g. gold. There are not special requirements in sample preparation. However, low S/N ratio is typical for sample with low refractive index and hence the application of this technique strongly depends on the type of a sample.

Since the numerical aperture NA is defined as $NA = n\sin\theta$, where n is the refractive index of the medium, in which the optics are surrounded and θ is the half-angle of the light cone that can enter or exit the optical element, the spatial resolution of FT-IR imaging can be improved by making contact with the sample with a material with a high refractive index like for attenuated total reflection (ATR) spectroscopy. In ATR FT-IR microscopy, a single-reflection hemispherical internal reflection element (IRE) is used, for instance a germanium (Ge) crystal. Since it shows a large refractive index ($n = 4$), the spatial resolution increases four times in comparison to transmission and transflection modes, where n for air is equal to 1. In addition, the angle of incidence (θ) increases from 30° to 50°, resulting in a further increase of spatial resolution. The schematic showing how

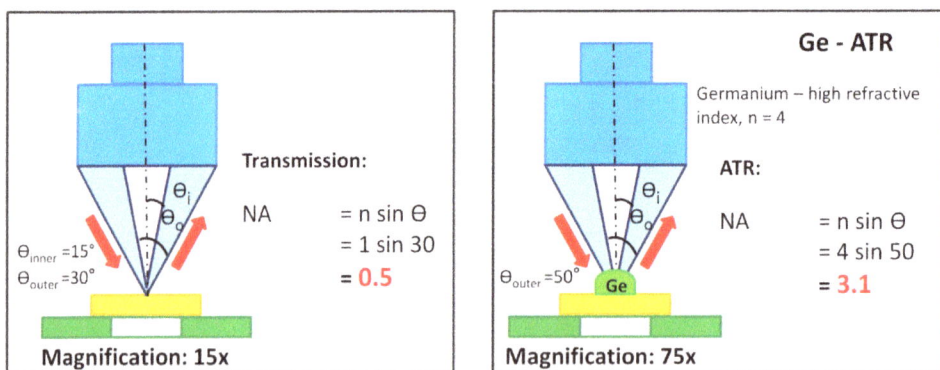

Fig. 3.2.3. Schematic diagram showing how the IRE element increases the spatial resolution in ATR FT-IR microscopy with the use of Ge ATR accessory. Illustration of courtesy of Dr. M. Kansiz and Agilent Technologies

the spatial resolution is increased due to the contact with the IRE is shown in Figure 3.2.3, while the comparison of chemical images recorded using transmission and ATR techniques and constructed for four-layer laminate film is shown in Figure 3.2.4.

Typically, the contact area between the ATR crystal and the sample is *ca.* 100×100 µm² in single spectrum measurements so if the sample is chemically heterogeneous in such an area, ATR FT-IR spectrum represents an overall chemical composition. In the FPA imaging, the use of the IRE and ATR mode provides chemical images of *ca.* 70×70 µm² size with max. spatial resolution of 1.1 mm at 4000 cm⁻¹. The disadvantage of the ATR FT-IR technique is the necessity for good contact between sample and ATR crystal that can distort or damage soft samples. As mentioned in Chapter 3.1. ATR FT-IR spectrum exhibits chemical information from a certain depth of penetration for a given IRE, which is rather small and does not exceed a few mm.

FT-IR microscopy and imaging are employed in several applications, which require the correlation between detection of a component and its spatial distribution in the sample. This includes for instance polymer science, forensic chemistry, art conservation and medical diagnostics. Data collection is fast in comparison to Raman imaging and takes few minutes. Chapters 6.10 and 6.11 present examples of the application of FT-IR imaging. Construction of IR images and data analysis is performed by using univariate and multivariate analysis as discussed in Chapter 5.

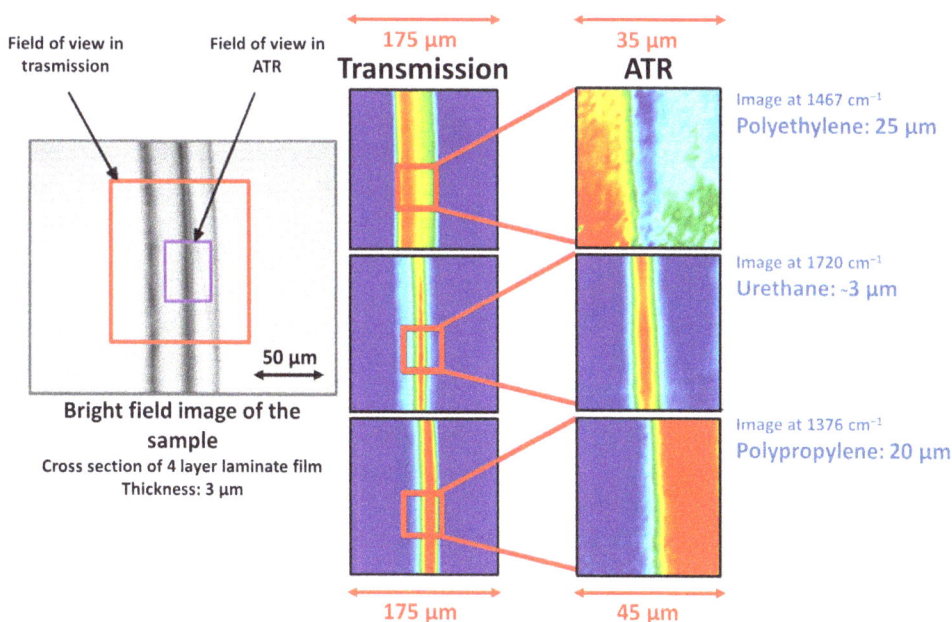

Fig. 3.2.4. Bright field image of a laminate film showing region of interest measured in the transmission and tranflection modes (*left*). Chemical images constructed by integration of IR bands specific for components of the film. Illustration of courtesy of Dr. M. Kansiz and Agilent Technologies

3.3. Vibrational circular dichroism

Piotr F. J. Lipiński

3.3.1. Chirality

Molecules of not a small number of chemical compounds are chiral, *i.e.* non-super-posable on their mirror image [1]. Chiral molecules exist in two stereoisomeric forms called *enantiomers*. Enantiomers have almost all their physical properties identical. An important exception is their behaviour to polarized light. Chiral molecules are optically active and enantiomers react in an exactly opposite manner in the presence of the polarized light. The phenomenon allowed to develop several useful spectroscopic techniques that extend our analytical opportunities in the field of chiral chemistry.

The development of the field is mainly driven by the needs of the pharmaceutical industry. Enantiomers have most often different biological activity. For example, S(-)-propranolol (a drug used in the treatment of hypertension) is 100 time more potent than R(+)-propranolol. [2] In the case of ketamine, an intravenous anaesthetic, the S(+) form is not only much more potent, but also less toxic. The opposite R(-) enantiomer causes agitation and hallucinations in the patients. The most famous example of chirality importance is probably thalidomide. [3] The substance was introduced to the market in 1957 as a mild but effective sedative, particularly suited for use against nausea and morning sickness very often experienced by pregnant women. In the following years, mothers using the drug gave birth to more than 10.000 children with severe limb malformations. The tragedy was caused by the S(-)-isomer present in the drug – medicinally inactive but terribly toxic. Since then a lot of strict regulations and requirements were introduced in the drug registration process. Their considerable part is devoted to the evaluation of enantiomers pharmacological action, or enantiomeric purity at every stage of drug development and manufacturing. This is why the field of chiral analytical chemistry is so important from the point of view of modern pharmaceutical industry.

3.3.2. What is Vibrational Circular Dichroism?

One of the fastest developing analytical tools in the field of chiral chemistry is Vibrational Circular Dichroism (VCD). To get straight to the heart of the technique, let us decompose its name.

VCD is *vibrational*, because it uses light from the infrared energy region, and as in IR spectroscopy it observes vibrational transitions. VCD is *circular*, because the light applied is left- or right-circularly polarized. Finally, VCD is kind of a *dichroism*, because it records a difference in absorbance (ΔA) (3.3.1):

$$\Delta A = A_L - A_R, \tag{3.3.1}$$

where A_L and A_R stand for the absorbance of left- and right- circularly polarized light.

Achiral substances have A_L equal to A_R, thus $\Delta A = 0$ and no VCD signal can be observed. In the case of chiral molecules, ΔA is greater or smaller than 0 (note that VCD signal can be both positive and negative), and their enantiomers exhibit exactly opposite values of ΔA at each frequency. VCD spectra of two enantiomers have all bands identical but with opposite signs. (Fig. 3.3.1).

3.3.3. How is VCD measured?

Figure 3.3.2 contains a general scheme of a modern VCD setup. An FT-IR spectrometer provides a randomly polarized beam of light, which is first passed through an optical filter. It is then linearly polarized by a polarizer and subsequently modulated by the key element of a VCD setup: photoelastic modulator (PEM). PEM changes, at a given frequency, the polarization of light between left- and right circularly polarized. The modulator is able to convert linearly polarized light to circularly polarized since the photoelastic material that is contained in it, can change its optical properties when a pressure is applied. The beam passes through the studied sample and falls at an IR detector. IR spectrum is obtained by a low pass filtering, while the VCD signal is high pass filtered. It still must be demodulated at the PEM frequency by a lock-in amplifier.

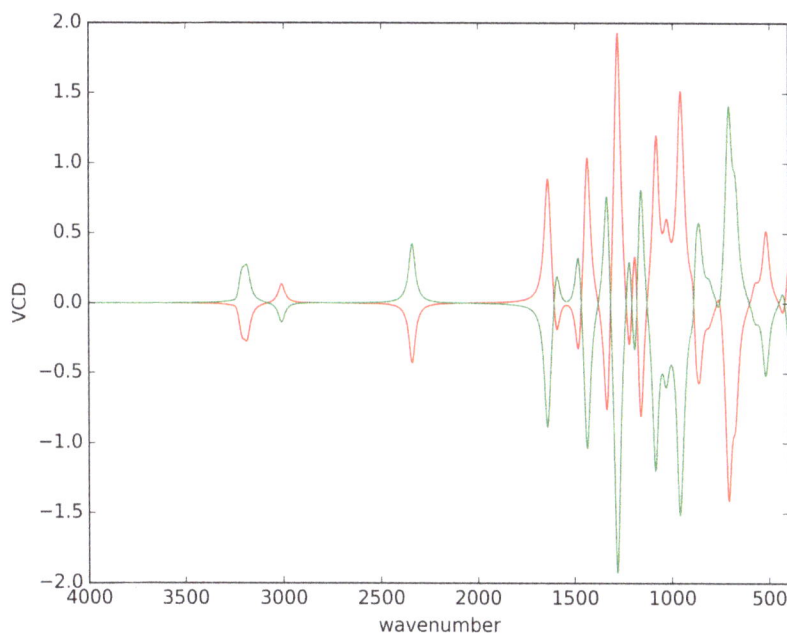

Fig. 3.3.1. Calculated VCD spectra of R (red) and S (green) enantiomers of 5-bromo-1-cyanoind(1H)-ene

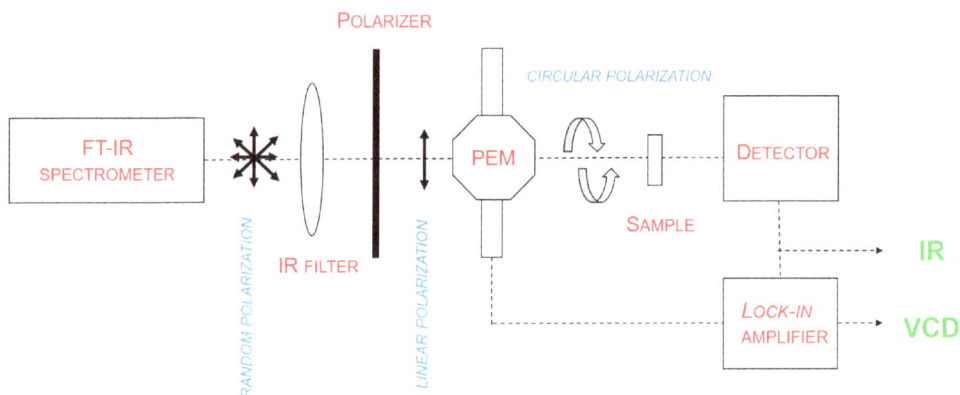

POLARIZER

CIRCULAR POLARIZATION

FT-IR SPECTROMETER

PEM

DETECTOR

IR FILTER

SAMPLE

RANDOM POLARIZATION

LINEAR POLARIZATION

IR

LOCK-IN AMPLIFIER

VCD

Fig. 3.3.2. Scheme of a typical modern VCD setup

Most measurements are made in the spectral range 2000 – 800 cm^{-1}, although these days some spectrometers are able to work in the range reaching 4000 cm^{-1}. Solvents used in VCD are deuterated IR transmitting solvents.

3.3.4. Applications of VCD spectroscopy

VCD has found several applications in chiral analysis:

1) determination of absolute configuration [4] (*which enantiomer is present in the sample?*);
2) determination of enantiomeric purity [5,6] such as methods development, along with identification and characterization of intermediates and impurities, are critical in the development of a chemical process. The preparation of a drug substance requires the development of analytical methods for monitoring reactions and identifying impurities. Methods development for a chiral drug molecule is more difficult as the method must be capable of monitoring the overall reaction as well as possible racemization of starting materials and products. Chiral methods are often required to monitor the reaction steps of a synthesis, however, the development of enantiomeric purity methods are time-consuming and expensive. The use of chiroptical detectors, such as circular dichroism (CD (*is the sample stereochemically pure or are both enantiomers present? if so, in what proportion?*);
3) monitoring of chiral reactions [7] which allows for VCD measurements even in the presence of strongly absorbing backgrounds. Time-dependent VCD spectra were analyzed by singular value decomposition and global exponential fitting. Spectral features corresponding to the complex and free (- (*when should I stop the reaction to get the best possible enantiomeric purity of the products?*);
4) exploring the structure and conformational behaviour of chiral compounds in the solution (*which conformers are present in the solution? what are their populations?*)

or intermolecular interactions in hydrogen-bonded as well as electron donor-acceptor complexes [8–11]

5) exploring the structure and conformation of proteins and peptides [12,13] (*which secondary structures are present?*).

The first of them is probably the most important and mature. The determination of absolute configuration is a vivid problem for pharmaceutical industry. VCD has a good chance of becoming a method of choice for configuration assignments. [4] It requires no crystallisation (as X-ray diffraction does) and no chiral shift reagents or derivatization have to be used (as in NMR). The calculations indispensable (see below) for the interpretation of VCD spectra are much simpler compared to simulation of optical rotation or electronic circular dichroism (ECD, circular dichroism in the UV-VIS region). Also, a molecule studied by VCD does not have to possess a chromophore (as in ECD).

The possible use of VCD in stereochemical assignments results from the fact that a pair of enantiomers has their VCD spectra identical but with opposite signs. A typical workflow to determine the configuration is to:

1. Measure a VCD spectrum of a sample.
2. Simulate a VCD spectrum of one of the enantiomers by quantum chemical calculations. A theoretical spectrum is necessary since there are no empirical rules that would say that, for example, a positive sign of a band is present in S or R enantiomer.
3. Compare the experimental and theoretical spectra. If the signs of the most important bands are identical, then the enantiomer in the sample is the one for which the simulation was performed. Otherwise, it is the opposite one.

3.3.5. Calculations of VCD spectra and their problems

It is then seen that the key to successful determination of absolute configuration by VCD is in the correctly simulated VCD spectrum. The theory and mathematics that lie behind the simulations is described elsewhere (shortly [14] or in detail [15]). For now, we should realize that three spectral parameters of a band are important here: frequency, intensity and sign (the fourth one, the band width, cannot be predicted by standard quantum chemical calculations). VCD frequencies are identical to those found in IR spectroscopy (connected with the force constant of a given vibrational mode). The intensity (I) is proportional to the product of the values of electric dipole transition moment (EDTM) and magnetic dipole transition moment (MDTM) vectors and the cosine of the angle ξ between the EDTM and MDTM vectors (3.3.2).

$$I \sim |EDTM||MDTM|\cos(\xi). \qquad (3.3.2)$$

The sign of the intensity is then clearly dependent on the angle ξ.

In many cases the simulations reproduce very well (or well enough to be of use) the experimental spectra. Still, there are however non-trivial problems that can complicate or misguide the determination of the absolute configuration.

First of all, most of the molecules studied in real-world problems are flexible and exist in several (several tens, hundreds etc.) conformers. Proper simulation of VCD spectra for such molecules requires conformational analysis (sometimes very time-consuming), calculation of the conformer populations (fraught with the risk of error), calculation of VCD spectra for all important conformers and averaging them over the Boltzmann populations.

If a molecule interacts with solvent (solvent effect) the frequency, intensity and sign of important bands can change. These changes may arise due to dielectric effect of the solvent, intramolecular hydrogen bonding, formation of dimers, hydrogen bonding between the molecules of substance and the solvent. [16] Sometimes it may not be easy to properly render these changes by QM calculations (especially, if the solvent is a hydrogen-bonding one as water, DMSO etc.). A successful simulation of a VCD spectrum may require in some cases very careful and time-consuming calculations.

3.3.5. Very short summary of current developments

The basic problems mentioned above: effect of conformation, solvent effects, effect of complexation as well as others like: chirality transfer, [8–11] substituent effect [17] or the interrelation between quantitative chirality and the VCD spectra [18] are a lively field of research.

Recent years have also witnessed interesting steps forward in the experimental settings that can enable wider application of VCD in chiral analysis. These include 2D-VCD, [19] transient vibrational chiral spectrometer [20] or measurements of VCD spectra with a tunable external-cavity quantum cascade. [21]

References

1. *Chirality*, IUPAC. Compend. Chem. Terminol. XML on-Line Corrected Version Http// goldbook.iupac.org, (1997).
2. Stoschitzky K., Lindner W., Rath M., Leitner C., Uray G., Zernig aGerald, Moshammer T., Klein W., *Stereoselective hemodynamic effects of (R)-and (S)-propranolol in man*, Naunyn. Schmiedebergs. Arch. Pharmacol., **339**, 474 (1989).
3. Eriksson T., Björkman S., Höglund P., *Clinical pharmacology of thalidomide*, Eur. J. Clin. Pharmacol., **57**, 365 (2001).
4. Kuppens T., Bultinck P., Langenaeker W., *Determination of absolute configuration via vibrational circular dichroism*, Drug Discov. Today Technol., **1**, 269 (2004).
5. Shah R.D., Nafie L.A., *Spectroscopic methods for determining enantiomeric purity and absolute configuration in chiral pharmaceutical molecules.*, Curr. Opin. Drug Discov. Devel., **4**, 764 (2001).

6. Urbanova M., Setnicka V., Volka K., *Measurements of concentration dependence and enantiomeric purity of terpene solutions as a test of a new commercial VCD spectrometer*, Chirality, **12**, 199 (2000).

7. Rüther A., Pfeifer M., Lórenz-Fonfría V.A., Lüdeke S., *Reaction Monitoring Using Mid-Infrared Laser-Based Vibrational Circular Dichroism.*, Chirality, **26**, 490 (2014).

8. Sadlej J., Dobrowolski J.Cz., Rode J.E., *VCD spectroscopy as a novel probe for chirality transfer in molecular interactions.*, Chem. Soc. Rev., **39**, 1478 (2010).

9. Rode J.E., Jamróz M.H., Dobrowolski J.Cz., Sadlej J., *On vibrational circular dichroism chirality transfer in electron donor-acceptor complexes: a prediction for the quinine···BF3 system.*, J. Phys. Chem. A, **116**, 7916 (2012).

10. Merten C., Xu Y., *Chirality transfer in a methyl lactate-ammonia complex observed by matrix-isolation vibrational circular dichroism spectroscopy.*, Angew. Chem. Int. Ed. Engl., **52**, 2073 (2013).

11. Merten C., Berger C.J., McDonald R., Xu Y., *Evidence of dihydrogen bonding of a chiral amine-borane complex in solution by VCD spectroscopy.*, Angew. Chem. Int. Ed. Engl., **53**, 9940 (2014).

12. Yasui S.C., Keiderling T.A., *Vibrational circular dichroism of polypeptides and proteins*, Mikrochim. Acta, **95**, 325 (1988).

13. Keiderling T.A., *Protein and peptide secondary structure and conformational determination with vibrational circular dichroism*, Curr. Opin. Chem. Biol., **6**, 682 (2002).

14. Dobrowolski J.C., Lipiński P.F.J., Rode J.E., Sadlej J., α-Amino Acids In Water: A Review Of VCD And ROA Spectra, in: Optical Spectroscopy and Computational Methods in Biology and Medicine, M. Barańska (Ed.), Springer Netherlands, Dordrecht, 2013: pp. 83–160.

15. Magyarfalvi G., Tarczay G., Vass E., *Vibrational circular dichroism*, Wiley Interdiscip. Rev. Comput. Mol. Sci., **1**, 403 (2011).

16. Polavarapu P.L., *Molecular structure determination using chiroptical spectroscopy: where we may go wrong?*, Chirality, **24**, 909 (2012).

17. Lipiński P.F.J., Dobrowolski J.Cz., *Substituent effect in theoretical VCD spectra*, RSC Adv., **4**, 27062 (2014).

18. Lipiński P.F.J., Dobrowolski J.Cz., *Local chirality measures in QSPR: IR and VCD spectroscopy.*, RSC Adv., **4**, 47047 (2014).

19. Ma S., Freedman T.B., Cao X., Nafie L.A., *Two-dimensional vibrational circular dichroism correlation spectroscopy: pH-induced spectral changes in l-alanine*, J. Mol. Struct., **799**, 226 (2006).

20. Bonmarin M., Helbing J., *A picosecond time-resolved vibrational circular dichroism spectrometer*, Opt. Lett., **33**, 2086 (2008).

21. Lüdeke S., Pfeifer M., Fischer P., *Quantum-cascade laser-based vibrational circular dichroism.*, J. Am. Chem. Soc., **133**, 5704 (2011).

4

Special Raman techniques

4.1. Resonance Raman scattering spectroscopy

Katarzyna M. Marzec, Jakub Dybaś

4.1.1. Resonance *versus* normal Raman scattering and fluorescence

Similar to normal Raman scattering (NR) Resonance Raman Scattering (RRS) can also be described as the two single-photon process. However, the difference appears in the energy of the excitation line. In the case of NR, such energy allows for the transition of the photon to the virtual state, which is far below the first electronic state. When the energy of virtual state corresponds to the energy of electronic excited state of a specific chromophoric group(s) in a molecule, the resonant enhancement is observed (Fig. 4.1.1.). The second part of the process is the same as in the case of NR: emission of the photon with the same (Rayleigh), lower (Stokes) or higher energy (anti-Stokes). The presence of the additional electronic transition in the case of RRS causes the strong enhancement (by a factor of 10^3-10^6) of specific bands originating from the chromophore group in Raman spectrum [1].

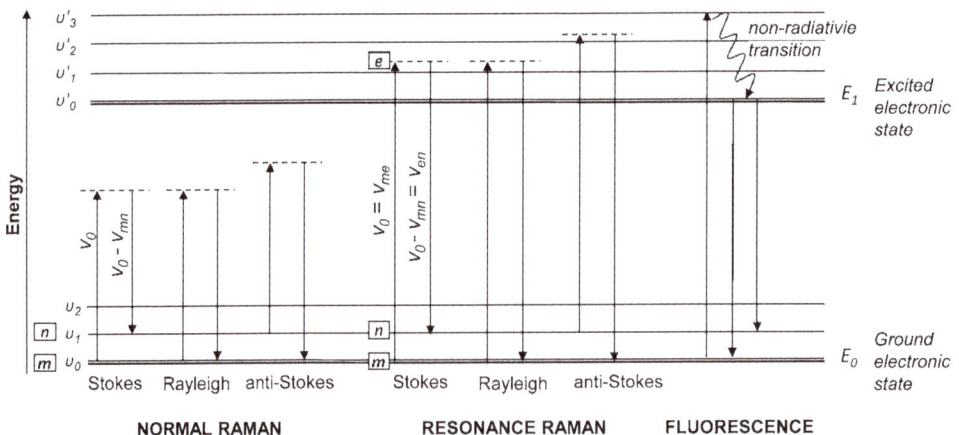

Fig. 4.1.1. Comparison of a simple diatomic energy levels for the normal Raman, resonance Raman and fluorescence spectra

The pre-resonance effect, which corresponds to the situation when the exciting line is close enough to the electronic excited state and also leads to bands enhancement, could also be observed.

The difference between RRS and fluorescence can be seen on the level of excited electronic state. The lifetime of this excited state for RRS is around 10^{-14} s, while for fluorescence, it may vary between 10^{-8}-10^{-5} s. Moreover, the fluorescence spectrum is observed when the excited state molecule decays via non-radiative transitions (vibrational relaxation) from the discrete vibrational level of the excited electronic state to the lowest vibrational level of the excited state (which is not observed in RRS). Subsequently, this process is followed by the emission of radiation. Weak NR signals may be overwhelmed by fluorescence signals, as fluorescence is characterized by a longer excited state lifetime. This situation is observed not only for fluorescent molecules excited with specific wavelengths (in the range of visible light), but also for many complex samples where the signal is coming from the components' matrix [2]. As an example we can present the autofluorescence of elastic lamina fibers when radiated with a laser wavelength of 532nm, even though the main components of this aorta structure (elastin and collagen) are not fluorescent molecules at this wavelength [3].

To eliminate the fluorescence interference in Raman spectra, different procedures may be carried out, starting with the use of different laser wavelengths as an excitation source. To obtain the Raman spectrum of some fluorescent proteins, a laser may be used to irradiate samples for some time before Raman measurement in order to cause the photon-induced destruction of the chromophore. Such a phenomenon is known as photobleaching and was previously used to obtain Raman spectra of proteins or biological samples. Fluorescence effects may also be reduced with the use of confocal Raman systems. In such laser scanning confocal instruments samples are penetrated only in a specific plane (the signal is not collected from the whole volume of the sample), reducing the fluorescent signal from potential contaminations. Secondly, in such conditions, the sample is excited to a high enough point to reach fluorophore saturation (molecules are in the excited state). As a consequence, an increase in the excitation wavelength produces an increase of Raman signal and a reduction in fluorescence emission.

4.1.2. Phenomenon of Resonance Raman scattering

As previously described in Chapter 2, the NR transition moment must have values different from zero (formula 2.3), which is determined by the change of polarizability during the transition between vibrational states, from initial m to final n (see Fig. 4.1.1.). Moreover, it was also postulated that the intensity of a normal Raman band is given by the equation 2.4, where $v_{mn} = v_{osc}$.

$$I_{mn} \propto I_0 \cdot (v_0 - v_{mn})^4 \cdot |\alpha_{mn}|^2. \tag{4.1}$$

In the case of RRS, α_{mn} will represent the change of polarizability α during the transition between the $m \rightarrow e \rightarrow n$ states, where e represents the electronic excited state (see Fig. 4.1.1.). That is why, the polarizability tensor α_{mn} in RRS depends on the frequencies (v_{me} and v_{en}), as well as on the electric transition dipole moments (M_{me} and M_{en}) which correspond to the energy differences between $m \rightarrow e \rightarrow n$ states.

In NR, the sample is irradiated with an exciting line which energy is much smaller than that of electronic transition, so $v_0 \ll v_{me}$. Contrary to NR, in RRS, v_0 approaches v_{me}, which also has an impact on the increase of the α_{mn} value, and consequently on the significant increase of the intensity (I_{mn}) of the Raman band at $v_0 - v_{mn}$. The intensity of resonance Raman scattering can be expected to be orders of magnitude greater than normal Raman scattering when v_0 approaches v_{me}.

This shows that compared to non-resonant NR, even components at low concentrations may be detected and analysed with the use of the proper excitation wavelength, which proves the high sensitivity of this technique. Using RRS, it is possible to analyse samples even with nanomolar concentrations [4]. This also explains that to properly understand the observed RRS profile of a sample it is useful to know the UV-Vis absorption spectrum of the sample. To obtain RRS, a given sample is irradiated with an exciting line which coincides with the wavelength of an electronic transition of the sample. That is why the UV-Vis profile allows us to choose an exciting line which corresponds to the electronic transition of specific sample chromophore.

Let's take the theoretical molecule X, containing two chromophoric groups A and B, which has an absorption spectrum with two maxima at wavelengths λ_A and λ_B. The theoretical model of UV-Vis spectrum of molecule X is presented in Fig. 4.1.2. To selectively enhance the vibrations of the chromophore A in a complex spectrum

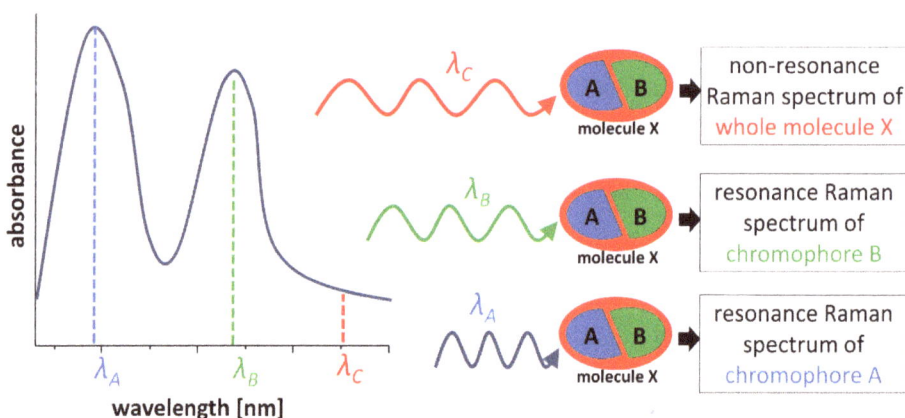

Fig. 4.1.2. The model of the UV-Vis absorption spectrum of molecule X containing two chromophoric groups A and B with two major absorption bands with the maxima at λ_A and λ_B. Exciting the Raman spectra at λ_A nm and λ_B nm results in two different resonantly enhanced Raman spectra of chromophore A and B, respectively. All parts of the molecule contribute to the non-resonant Raman spectrum excited at λ_C nm

of the molecule, the exciting line has to have ν_0 near to ν_A (which corresponds properly to λ_A). Vibrations of chromophore B will be enhanced when the laser wavelength will be equal to λ_B. On the other hand, if the exciting line λ_C is used, the non-resonant Raman spectrum which comes from all parts of molecule will be observed.

For this reason, the UV–Vis electronic absorption spectrum of the studied compound, which shows the allowed electronic transitions will help us to choose the best laser wavelength in order to observe the vibrations of specific chromophore. By changing the excitation wavelength, the different RRS of the same molecule may be obtained and give us information about specific parts of the molecule. Such selective enhancement suggests a high specificity of the RRS technique.

The detailed information about UV-Vis spectrum may provide additional information about the RRS origin. The quantitative description of RRS scattering theory was provided by Albrecht et al., who showed how RRS intensity can arise from several mechanisms, mainly from A-term (Franck–Condon) and B-term (Herzberg––Teller vibronic coupling) [5-7].

In type A or Franck–Condon scattering, only totally symmetric modes are enhanced in RRS. Such mechanism is observed for many compounds. As an example we can include here the RR spectra of TiI_4 and NH_3 obtained by 514.5 and 216.8 nm excitations, respectively [1]. Non–Condon dependence of the electronic transition moment upon the vibrational coordinate is possible in B–term enhancement, where both symmetric and non-symmetric fundamentals can be enhanced. However, the magnitude of B-term enhancement of symmetric vibrations is lower than that of A-term enhancement. B-term enhancement will dominate only for non-symmetric vibrations [8]. Such RR scattering [9] involves vibronic coupling between the two allowed excited electronic transitions. This mechanism is observed for metalloproteins being excited with the laser wavelength which corresponds to the electronic transitions of the Q band of the UV-Vis spectrum. If the enhancement of fundamentals cannot occur via A or B-terms, as transition is rigorously forbidden at the equilibrium geometry, then C-term enhancement of overtones and combinations modes may occur [8,10].

4.1.3. Application and potential of RRS

As we proved above, RRS is characterized by high sensitivity and selectivity in comparison with NR, which gives RRS technique the advantage in many analytical studies. Similar to NR, RRS allows for the study of samples in the gaseous, liquid and solid state.

In art history, archaeology and forensics RRS is successfully used to study the composition of different pigments and dyes. It is also known as a non-invasive and non-destructive method of assessing the distribution and concentration of various biomolecules inside plant and animal tissues. It has been applied to study the carotenoid status in human skin, as a biomarker of fruit/vegetable intake [11]. A single RRS skin measure allowed for the classification of inter-individual variability in skin carotenoid status and to identify factors associated with the biomarker in this population [12]. It is also possible to differentiate various retinoid fractions from a mixture with the use of this technique utilizing different excitation wavelengths. Upon excitations with different wavelengths it was possible to differentiate lutein, violaxanthin, β-carotene and 9-cis neoxanthin [13]. The use of the 532 nm excitation laser line, which allows observation of the pre-resonance Raman spectrum of retinols, was also used to study the distribution of vitamin A component in liver and lung tissues [14-15].

Resonance Raman spectroscopy has long been applied to monitor the molecular dynamics of different metalloproteins, among which the most common is hemoglobin [16]. This highly symmetrical and chromophoric heme prosthetic group provides strong resonance enhancement, especially when the excitation wavelength is in resonance with the intense electronic transitions cantered at ~400 nm (Soret), 525 nm (Q_v or α band) and 575 nm (Q_0 or β band) [17]. Moreover, peptide chains of heme proteins may also be studied with this technique as they exhibit transitions below 250 nm. RRS was successfully used not only for standard hemoporphirins, but also for the detection, analysis and visualization of 2D and 3D distributions of heme in both cells and tissues [18,19]. RRS provides excellent signal–to–noise ratio spectra with very high reproducibility from single erythrocytes, which allows for the study of various hemopathies [20].

As in case of heme proteins, peptide chains of other proteins, or protein–drug interactions, may also be studied with the use of excitation sources below 260 nm. The use of RRS in such deep UV was successfully applied in order to investigate DNA, RNA and nucleic acid components [21,22]. The use of RRS in such deep UV is mainly used in the bioanalytical and life science fields, however it is also useful to study solid catalysts and heterogeneous catalytic reactions [23].

Because of the effects of this vibrational technique, information about the electronic structure of a studied sample can be obtained. This makes RRS a very useful technique in nanotechnology and materials science in order to study and characterize structures of such materials as carbon nanotubes, graphite, graphene and others [24,25].

4.1.4. Instrumentation

As already presented and described in Chapter 2, for RRS detection the standard Raman instrumentation may be applied. As mentioned before, selective resonance Raman enhancement of specific chromophores of molecules may be obtained by changing the excitation wavelength. That is why tunable lasers, in which the wavelength can be altered within a specific range, are commonly applied to this technique. To provide positive identification, even with higher than RRS sensitivity and selectivity, RRS is successfully used in combination with liquid chromatography [26] and SERS (surface-enhanced resonance Raman scattering, SERRS). Tip-enhanced Raman spectroscopy (TERS), which is a variation of SERS, may also use the resonance effect (TERRS) and is a promising technique for future nanoanalysis [27].

References

1. Ferraro J.R., Nakamoto K., *Introductory Raman Spectroscopy*, Academic Press Inc., San Diego 1994.
2. Matousek P., Towrie M., Parker A.W., *Fluorescence background suppression in Raman spectroscopy using combined Kerr gated and shifted excitation Raman difference techniques*, J. Raman Spectrosc., **33**, 238 (2002).
3. Marzec K.M., Wróbel T.P., Ryguła A., Maslak E., Jasztal A., Fedorowicz A., Chlopicki S., Barańska M., *Visualization of the biochemical markers of atherosclerotic plaque with the use of Raman, IR and AFM*, J. Biophotonics, **7**, 744 (2014).
4. Krishnan R.S., Shankar R.K., *Raman effect: History of the discovery*, J. Raman Spectrosc., **10**, 1 (1981).
5. Albrecht A.C., *On the theory of Raman intensities*, J. Chem. Phys., **34**, 1476 (1961).
6. Albrecht A.C. and Hutley M.C., *On the Dependence of Vibrational Raman. Intensity on the Wavelength of Incident Light*, J. Chem. Phys., **55**, 4438 (1971).
7. Tang J., Albrecht A.C., *Raman Spectroscopy: Theory and Practice* (eds. H. A. Szymanski) Plenum, New York 1970, 2, pp. 33–68.
8. Asher S.A., *UV resonance Raman studies of molecular structure and dynamics: applications in physical and biophysical chemistry*, Anun. Rev. Phys. Chem., **39**, 537 (1988).
9. Wang J., Takahashi S., Rousseau D.L., *Identification of the overtone of the Fe–CO stretching mode in heme proteins: a probe of the heme active site*, Proc. Natl. Acad. Sci., **92**, 9402 (1995).
10. Marzec K.M., Perez–Guaita D., De Veij M., McNaughton D., Barańska M., Dixon M.W.A., Tilley L., Wood B.R., *Red Blood Cells Polarize Green Laser Light Revealing Hemoglobin's Enhanced Non–Fundamental Raman Modes*, Chem. Phys. Chem., **15**, 3963 (2014).
11. Scarmo S., Cartmel B., Lin H., Leffell D.J., Ermakov I.V., Gellermann W., Bernstein P.S., Mayne S.T., *Single v. multiple measures of skin carotenoids by resonance Raman spectroscopy as a biomarker of usual carotenoid status*, Br. J. Nutr., **110**, 911 (2013).
12. Scarmo S., Henebery K., Peracchio H., Cartmel B., Lin H., Ermakov I.V., Gellermann W., Bernstein P.S., Duffy V.B., Mayne S.T., *Skin carotenoid status measured by resonance Raman spectroscopy as a biomarker of fruit and vegetable intake in preschool children*, Eur. J. Clin. Nutr., **66**, 555 (2012).
13. Andreeva A., Velitchkova M., *Resonance Raman spectroscopy of carotenoids in Photosystem I particles*, Biophysical Chemistry, **114**, 129 (2005).
14. Kochan K., Marzec K.M., Chruszcz–Lipska K., Jasztal A., Maslak E., Musiolik H., Chlopicki S., Barańska M., *Pathological changes in the biochemical profile of the liver in atherosclerosis and diabetes assessed by Raman spectroscopy*, Analyst, **138**, 3885 (2013).
15. Marzec K.M., Kochan K., Fedorowicz A., Jasztal A., Chruszcz–Lipska K., Dobrowolski J. Cz., Chlopicki S., Barańska M., *Raman microimaging of murine lungs: insight into the vitamin A content*, Analyst, **140**, 2171 (2015).
16. Spiro T.G., *Biological Applications of Raman Spectroscopy: Resonance Raman Spectra of Heme and Metalloproteins Vol. 3* John Wiley & Sons, New York 1988.

17. Yamamoto T., Palmer G., *The valence and spin state of iron in oxyhemoglobin as inferred from resonance Raman spectroscopy*, J. Biol. Chem., **248**, 5211 (1973).
18. Marzec K.M., Ryguła A., Wood B.R., Chlopicki S., Barańska M., *High-resolution Raman imaging reveals spatial location of heme oxidation sites in single red blood cells of dried smears*, J. Raman Spectrosc., **46**, 76 (2015).
19. Wood B.R., Caspers P., Puppels G.J., Pandiancherri S., McNaughton D., *Resonance Raman spectroscopy of red blood cells using near–infrared laser excitation*, Anal. Bioanal. Chem., **387**, 1691 (2007).
20. Wood B.R., McNaughton D., *Vibrational Spectroscopy for Medical Diagnosis*, (eds. M. Diem, P. R. Griffiths, J.M. Chalmers) John Wiley & Sons, UK 2008, pp. 261–309.
21. Blazej D.C., Peticolas W.L., *Ultraviolet resonant Raman Spectroscopy ofnucleicacidcomponents*, Proc. Nati. Acad. Sci., **74**, 2639 (1977).
22. Wojtuszewski K., Mukerji I., *The HU–DNA binding interaction probed with UV resonance Raman spectroscopy: Structural elements of specificity*, Protein Sci., **13**, 2416 (2004).
23. Kim H., Kosuda K.M., Van Duyne K.P., Stair P.C., *Resonance Raman and surface– and tip–enhanced Raman spectroscopy methods to study solid catalysts and heterogeneous catalytic reactions*, Chem. Soc. Rev., **39**, 4820 (2010).
24. Jorio A., Pimenta M.A., Souza Filho A.G., Saito R., Dresselhaus G., Dresselhaus M.S., *Characterizing carbon nanotube samples with resonance Raman scattering*, New. Phys., **5**, 139.1 (2003).
25. Zolyomi V., Koltai J. and Kurti J., *Resonance Raman spectroscopy of graphite and graphene*, Phys. Status Solidi B, **248**, 2435 (2011).
26. Dijkstra R.J., Ariese F., Gooijer C., Brinkman U.A.Th., *Raman spectroscopy as a detection method for liquid–separation techniques*, TrAC, **24**, 304 (2005).
27. Taguchi A., Hayazawa N., Furusawa K., Ishitobi H., Kawata S., *Deep–UV tip–enhanced Raman scattering*, J. Raman Spectrosc., **40**, 1324 (2009).

4.2. Surface-enhanced Raman scattering spectroscopy (SERS)

Agata Królikowska, Jolanta Bukowska

Inelastic scattering of light, called Raman effect, is a weak phenomenon: only 1 out of 10^7 photons is scattered at a frequency different from the incident photons. Remaining fraction of the radiation is scattered elastically (Rayleigh scattering). Together with development of instrumentation (light sources, ultrasensitive detectors) a large progress in quality of the collected Raman spectra has emerged. Still an enhancement of the Raman scattered light is essential in many experiments, particularly when one wants to probe few molecule events. Three main strategies of spectrum enhancement can be applied in Raman spectroscopy:

1. use of incident light of energy matching the energy of an allowed molecular electronic transition (resonance Raman effect, *see* Chapter 4.1),
2. exploiting so called non-linear Raman effects,
3. utilizing interactions of molecule with the surface on metal nanoparticles.

The third approach underlies the basis of *surface-enhanced Raman scattering* spectroscopy, commonly abbreviated as SERS.

SERS phenomenon was first observed by Fleischmann and co-workers in 1974 [1], who collected Raman signal for pyridine adsorbed on silver electrode, but not fully understood the significance of the discovery. These were two other groups of van Duyne [2] and Creighton [3], who proved in 1977 that recording of Raman spectrum for adsorbed molecules is actually possible due to a giant increase of the Raman scattering signal. It was demonstrated that the source of this enhancement, estimated to be in the order of $10^4 - 10^6$, is the surface of a rough Ag electrode. Extensive studies, both experimental and theoretical, carried out in next decades resulted in deeper understanding of physical fundamentals of SERS phenomenon and recognition of mechanism of Raman signal amplification for some types of metal surfaces.

4.2.1. Mechanism of surface enhancement

The main origin of an enhanced Raman scattering for chemical molecules located in the proximity to metal nanoparticles is a huge electromagnetic field generated by metallic nanoparticles. A phenomenon responsible for creating such a strong field is called plasmon resonance. Plasmon resonance occurs when a frequency of the electromagnetic radiation incident on the molecule is very close to a frequency of oscillations of electron gas in metal. For smooth metal surfaces frequencies corresponding to oscillations of electrons are quite different from typical frequencies used for excitation in Raman spectroscopy (visible light range, near infrared). However, if the surface is rough and atomic clusters of a size in the range of $10-10^2$ nm are present, then this light may excite so called *localized surface plasmons* (LSP), which are collective oscillations of the conduction band electrons, localized within the metal nanostructures. When the frequency of the laser used to excite Raman spectrum (field of intensity E_0) is close to the frequency of LSP, then surface plasmon resonance will occur, giving rise to a very strong electromagnetic field E_s by metallic nanoparticle (Fig. 4.2.1). Intensity of this field is determined by geometry

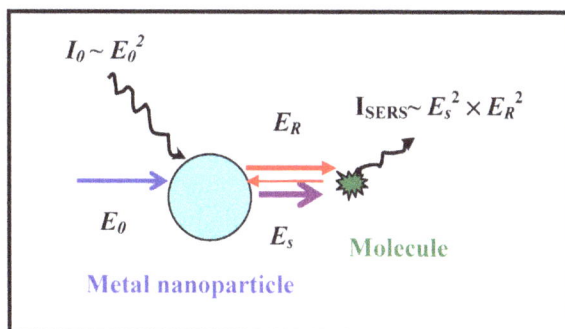

$$I_0 \sim E_0^2$$

$$I_{SERS} \sim E_s^2 \times E_R^2$$

E_R

E_0

E_s

Molecule

Metal nanoparticle

Fig. 4.2.1. Mechanism of electromagnetic enhancement (see the text for the symbol explanations)

and size of the nanostructure, type of the metal and electric permittivity of the surroundings. Raman scattering by the molecule adsorbed onto metal nanostructure is hence excited by already enhanced field of intensity E_s, generating a field E_R, which has a frequency of Raman scattering. Field E_R can be further enhanced by nanoparticle in a similar way like in the first step, namely a plasmon resonance.

As a result, recalling that intensity of the Raman scattered light is proportional to E^2, it can be seen that intensity of a given band in SERS spectrum is determined by a product of E_s^2 and E_R^2 (4.2.1):

$$I_{SERS} \sim E_s^2 \times E_R^2. \tag{4.2.1}$$

A surface plasmon resonance condition is certainly not equally met for both frequencies (frequency of the incoming light and the Raman scattered light frequency). However, the Raman shifted and exciting light frequencies are close to each other for the low frequency vibrational modes and thus it can be approximated that values of E_R and E_s are also similar. Hence, we can use an approximation (4.2.2):

$$I_{SERS} \sim E_s^4. \tag{4.2.2}$$

Equation (4.2.2) explains a giant increase of the Raman scattering intensity for the molecules adsorbed on metal nanoparticles.

Described above electromagnetic mechanism contributes the most to the surface enhancement phenomenon. On the other hand, several experimental observations, among them a fact that stronger enhancement is experienced by chemisorbed molecules on the surface than for weakly physisorbed ones, resulted in a search for other sources of the enhancement in SERS spectrum. It is assumed that beyond an electromagnetic enhancement, there is a contribution of a chemical effect, called also *charge transfer* (CT) effect (charge transfer between the metal and adsorbed molecule) to the total surface enhancement.

The phenomenon is considered as a two-step process. In a first step, a photon of the incident laser, having energy $h\nu_0$ drives a transfer of the electron from a Fermi level of the metal into LUMO (*lowest unoccupied molecular orbital*) level of the molecule adsorbed on metal nanoparticle. In a second step electron returns to the metal, which is accompanied by emission of a photon with the lower energy (Stokes photon), coupled with a simultaneous excitation of a molecular vibration. A charge transfer from the *highest occupied molecular orbital* (HOMO) of the molecule to the metal is also possible. In each case there must be a matching between the incident photon energy and the energy separation between Fermi level and LUMO or HOMO level. Therefore this process can be considered as similar to resonance Raman effect, but involving metallic energy level (within CT complex). Enhancement by CT effect is typically believed to be relatively small, lying in the range of 10-10^2. The overall enhancement factor (*EF*) is given by a product of electromagnetic (*EF_{EM}*) and chemical (EF$_{CT}$) contributions (4.2.3):

$$EF = EF_{EM} \times EF_{CT} \tag{4.2.3}$$

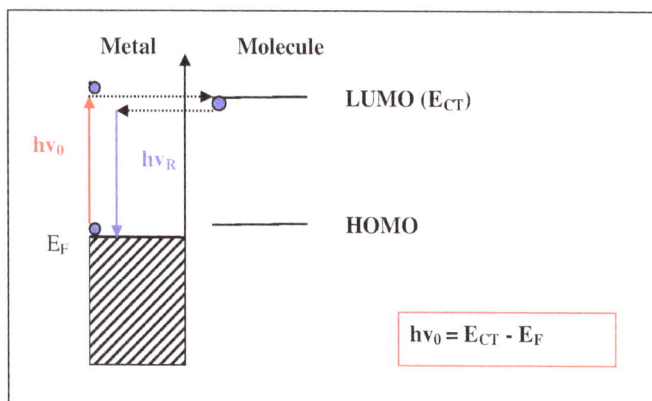

Fig 4.2.2. Enhancement by chemical effect (CT) for charge transfer from the metal to the molecule. E_F – energy of the Fermi level in the metal. Enhancement occurs when the energy of the exciting beam (hv_0) is matching the energy separation between LUMO level (E_{CT}) and E_F. hv_R – Raman photon

SERS spectrum can be additionally enhanced using excitation beam of energy lying in the range of electronic absorption of a given molecule. Under these conditions, as already known, a resonance Raman effect is observed in the absence of the metallic nanoparticles, resulting in a few orders of magnitude stronger Raman spectrum. If the molecule is in addition adsorbed on a nanostructured substrate, prepared in order to obtain a high surface enhancement for used excitation wavelength, then these two types of the enhancement: molecular (resonance effect) and surface (SERS) can be superimposed and considerably stronger spectrum will be obtained comparing to conventional SERS effect. Such phenomenon is called *surface-enhanced resonance Raman scattering* (SERRS).

4.2.2. Types of the substrates used in SERS spectroscopy

A key task in SERS spectroscopy is preparation of a high quality metallic substrate, providing high values of surface enhancement factor. Although many metals exhibit surface enhancement, the best results are obtained for silver, gold and copper. Susceptibility of Cu surface to oxidation makes silver and gold first choice materials for SERS substrates. Optical properties of these metals make SERS working range of Ag much larger than for Au. High surface enhancement can be obtained in a large range of laser radiation energy used in Raman spectroscopy (visible and near infrared) for Ag, while the enhancement in case of Au can be only observed for lower energetic radiation (red and near infrared). Nanostructures of an appropriate size and shape must be fabricated in order to get a satisfying enhancement of Raman spectrum by surface plasmon resonance. A list of nanostructures acting as efficient nanoresonators and supplying a large enhancement is very long. The most important types of the substrates employed for SERS spectroscopy include:

1. electrochemically roughened electrodes,
2. metal colloidal suspensions, e.g. synthesized through chemical reduction or by laser ablations; such prepared nanoparticles can be also deposited on solid substrates, not providing SERS activity (glass, Si, smooth gold),
3. smooth Ag or Au films deposited on polystyrene or silica nanospheres of controlled size, assembled on a solid substrate (FON – *film over nanospheres*),
4. periodically ordered nanostructures formed on solid substrates, e.g. formed using electron beams lithography.

First two methods are the most popular due to simplicity of preparation and effective enhancement of Raman spectrum. It is worth recalling, that plasmonic properties of the nanostructures and hence their electromagnetic enhancement factors can be successfully improved by controlling their geometry and size. It was demonstrated that the most intense fields, critical for the enhancement are generated, e.g. on the sharp edges and corners of the polyhedrons and in the small junctions (around 2 nm) between aggregated nanoparticles. For this reason triangular, cubic, star- or flower-shaped nanoparticles can be effective nanoresonators. An important factor is also an appropriate aggregation of nanoparticles, generating between them a region for which electromagnetic fields are highly concentrated. Typically this aim is accomplished by adsorption of chemical species on nanoparticles, providing junctions at a controlled distance of aggregates. In some experiments it is important to get the plasmon resonance at particular frequencies. This is the case for biological systems, which are strongly fluorescent. One must excite Raman spectrum using a beam in a near infrared region to avoid this problem. It is possible using spherical core-shell nanoparticles, composed of SiO_2 (core) and Ag or Au (shell). Similar effect can be achieved for hollow gold/silver nanospheres.

4.2.3. SERS spectral features

Both band positions and intensities in SERS spectrum are different than those in normal Raman spectrum of the same chemical, which can be ascribed both to a mechanism of surface enhancement and a fact that SERS signal corresponds to the molecules interacting with metal. SERS bands are quite often broadened. Fewer bands are also typically observed in SERS spectrum, as both electromagnetic and CT term of the surface enhancement may selectively enhance some vibrational modes.

Electric field on the metal surface is strongly anisotropic: normal component is significantly larger than the tangential one. Raman modes with the change of polarizability component normal to the surface will exhibit the strongest enhancement. Estimation of the average orientation of adsorbed molecules versus the metal surface from SERS spectrum is hence possible.

4.2.4. Applications of SERS spectroscopy

Numerous applications of SERS spectroscopy can be found. At present, sensitivity of this method can reach even *single molecule* level. Therefore SERS shows a great potential in analytics and bioanalytics. Nanosensors exploiting SERS effect for pH determination, quantitative detection of metal cations and toxic species have been developed. A glucose sensor utilizing SERS signal was patented. Efforts to use SERS spectroscopy as a diagnostic tool in oncology have been also made. A present review of both description of SERS phenomenon and its numerous applications can be found in the review papers [4,5].

References

1. Fleischmann M., Hendra P.J., McQuillan A.J., *Raman spectra of pyridine at a silver electrode*, Chem. Phys. Lett., **26**, 163 (1974).
2. Jeanmaire D.L., Van Duyne R.P., *Surface Raman Spectroelectrochemistry*, J. Electroanal. Chem., **84**, 1 (1977).
3. Albrecht M. G., Evans J.F., Creighton J.A., *Anomalously intense Raman spectra of pyridine at a silver electrode*, J. Am. Chem. Soc., **99**, 5215 (1977).
4. Bukowska J., Piotrowski P., in: *"Optical Spectroscopy and Computational Methods in Biology and Medicine"* (ed. M. Barańska), Springer, Germany 2014, pp. 29-59.
5. Schlücker S., *Surface-Enhanced Raman Spectroscopy: Concepts and Chemical Applications*, Angew. Chem. Int. Ed. **53**, 4756 (2014).

4.3. Raman Optical Activity (ROA)

Joanna E. Rode

Raman Optical Activity (ROA) is one of the chiroptical methods that measures the difference in intensity between Raman scattered right and left circularly polarized incident light. To explain the difference in the response of a molecule between right- and left-circularly polarized light, one needs to go beyond the electric dipole approximation and in addition to the electric dipole moment induced by the electric field of the optical wave, it is necessary to incorporate the electric dipole induced by the oscillating magnetic field and electric quadrupole moments. Only chiral molecules, *i.e.* exhibiting all types of chirality elements, i.e. center of chirality, chirality axis, helicity, chirality plane and supramolecular topological chirality are active in ROA and other chiroptical techniques. The theoretical background for the ROA phenomenon was first provided by Atkins and Barron in 1969 [1]. Two years later it was further developed by Barron and Buckingham [2], who introduced a dimensionless circular intensity difference (CID), defined as the intensity ratio of ROA to Raman scattering. CID is now denoted by Δ and called the incident circular polarization ICP-ROA (4.3.1):

$$CID = \Delta_\alpha = (I_\alpha^R - I_\alpha^L)/(I_\alpha^R + I_\alpha^L), \qquad (4.3.1)$$

where I_α^R and I_α^L are the scattered intensities with α-polarization in right- and left-circularly polarized (RCP and LCP, respectively) incident light. The Δ ratio amounts to less than 10^{-3}, thus the ROA signal is much weaker than the parent Raman one.

4.3.1. Schematic diagram of ROA phenomenon

Like in Raman spectroscopy, the inelastic ROA scattering is a two-photon process composed of the absorption of a photon into a virtual state and the subsequent immediate emission of a photon of different energy. Consequently, the molecule returns from the virtual state to the electronic ground but with a vibrationally changed level. Therefore, the process is called inelastic – the quantum energies of the incident and scattered light are different. The two-photon process gives rise to four different types of ROA measurements, corresponding to distinct polarizations of the incident and scattered beams (Fig. 4.3.1) [3]. The original form of ROA, described by the CID expression [2], is called the incident circular polarization ICP-ROA. In this case, the polarization of the incident laser beam is modulated between right- and left- directions. The Raman intensity is detected at a fixed linear or unpolarized radiation state (Fig. 4.3.1a).

The second form of ROA is called the scattered circular polarization SCP-ROA. In this form, the incident laser radiation is either unpolarized or linearly polarized and the difference in the right- and left- circularly polarized Raman light is measured for the scattered beam (Fig. 4.3.1b). In the far-from-resonance approximation, the intensity of the ICP-ROA and SCP-ROA bands are the same, however, the experimental setup is completely different [4]. The in-phase dual circular polarization (DCP$_I$) ROA is the third form of the ROA experiment. This time, the polarizations of both the incident and scattered radiation are switched synchronously between the right- and left- circular states (Fig. 4.3.1c). The fourth form of the ROA measurements is called the out-of-phase dual circular polarization (DCP$_{II}$) ROA, where the polarizations of both the incident and scattered radiation are switched oppositely between left- and right-circular states (Fig. 4.3.1d). The DCP$_I$ and DCP$_{II}$ -ROA predicted in 1989 [5] were only measured rarely [6, 7].

Since the light in ROA experiments is scattered in every direction, it can be detected at different angles with respect to incident beam. As in the Raman experiments, there are three preferred scattering geometries: the right-angle scattering ($\theta=90°$), the forward scattering ($\theta=0°$), and the backward scattering ($\theta=180°$), where θ is the angle between the incident and scattered radiation.

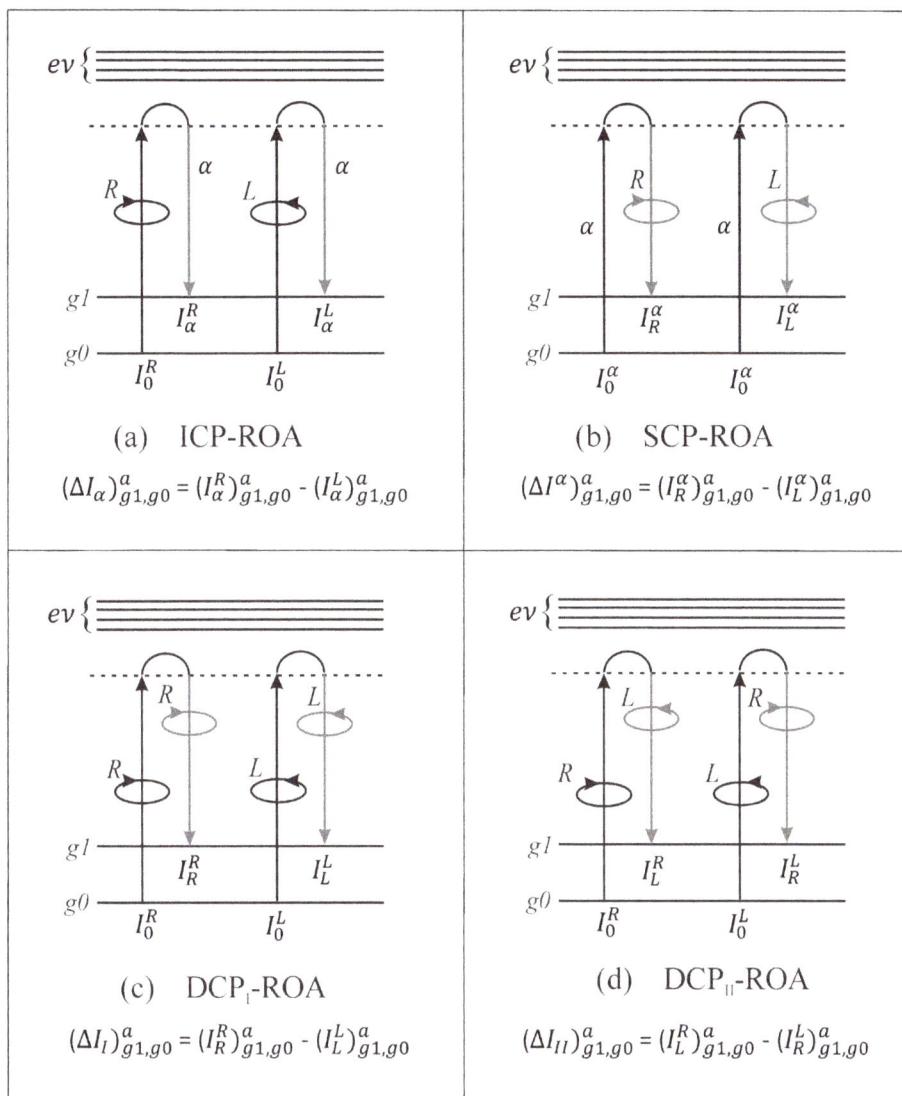

(a)   ICP-ROA

$$(\Delta I_\alpha)^a_{g1,g0} = (I^R_\alpha)^a_{g1,g0} - (I^L_\alpha)^a_{g1,g0}$$

(b)   SCP-ROA

$$(\Delta I^\alpha)^a_{g1,g0} = (I^\alpha_R)^a_{g1,g0} - (I^\alpha_L)^a_{g1,g0}$$

(c)   DCP$_{\mathrm{I}}$-ROA

$$(\Delta I_I)^a_{g1,g0} = (I^R_R)^a_{g1,g0} - (I^L_L)^a_{g1,g0}$$

(d)   DCP$_{\mathrm{II}}$-ROA

$$(\Delta I_{II})^a_{g1,g0} = (I^R_L)^a_{g1,g0} - (I^L_R)^a_{g1,g0}$$

Fig. 4.3.1. Energy-level diagrams illustrating different types of the ROA measurements. R, L, and α denote right, left, and none polarizations of the beams, the superscripts and subscripts refer to the incident and scattered light, respectively. $g0$ and $g1$ are the vibrational levels of the ground electronic state. The dotted line represents the virtual state whereas ev represents vibrational levels of the first excited state. (Reproduced from Ref. [3] with kind permission of John Wiley and Sons)

4.3.2. Theoretical description of ROA phenomenon

When considering the interaction of molecules with circularly polarized light, in addition to the interaction with the electric dipole (μ), we have also taken into account the contributions from the interactions of electric field of the optical wave

(E) with an oscillating magnetic dipole (m) and the electric quadrupole moment (Θ). For a more complete description it is also necessary to consider the electric dipole induced by an oscillating magnetic field (B) and electric field gradient (∇E) of the optical wave [8]. In the far-from-resonance approximation the expressions that describe the components of the electric dipole (μ_α), magnetic dipole (m_α) and electric quadrupole moment ($\Theta_{\alpha\beta}$) induced in the molecule by a real part of the electric vector (E_β), the magnetic vector ($\dot{B}_\beta$) and the gradient of the electric field of the incident light ($\nabla_\beta E_\gamma$) are as follows (4.3.2 – 4.3.4):

$$\mu_\alpha = \alpha_{\alpha\beta}E_\beta + \frac{1}{\omega}G'_{\alpha\beta}\dot{B}_\beta + \frac{1}{3}A_{\alpha\beta\gamma}\nabla_\beta E_\gamma + \cdots, \tag{4.3.2}$$

$$m_\alpha = -\frac{1}{\omega}G'_{\alpha\beta}\dot{E}_\beta + \cdots, \tag{4.3.3}$$

$$\Theta_{\alpha\beta} = A_{\alpha\beta\gamma}E_\gamma + \cdots, \tag{4.3.4}$$

where α, β, γ are spatial directions, $\alpha_{\alpha\beta}$ is the electric dipole – electric dipole molecular polarizability tensor responsible for light scattering and refraction; $G'_{\alpha\beta}$ is the electric dipole – magnetic dipole optical activity tensor whose isotropic part is responsible for optical rotation in fluids; and $A_{\alpha\beta\gamma}$ is the electric dipole – electric quadrupole optical activity tensor responsible for additional contributions to optical rotation in oriented samples.

Utilizing the expressions for the polarizability tensor $\alpha_{\alpha\beta}$ and activity tensors $G'_{\alpha\beta}$ and $A_{\alpha\beta\gamma}$ for a chiral molecule of a much smaller size than the wavelength of the incident light, fixed in the X, Y, Z coordinates, the ICP CID expressions can be written as follows (4.3.5 – 4.3.7) [9]:

$$\Delta(0°) = \frac{4[45\alpha G' + \beta(G')^2 - \beta(A)^2]}{c[45\alpha^2 + 7\beta(\alpha)^2]}, \tag{4.3.5}$$

$$\Delta_x(90°) = \frac{2[45\alpha G' + 7\beta(G')^2 + \beta(A)^2]}{c[45\alpha^2 + 7\beta(\alpha)^2]}, \tag{4.3.6a}$$

$$\Delta_z(90°) = \frac{12[\beta(G')^2 - \frac{1}{3}\beta(A)^2]}{6c\beta(\alpha)^2}, \tag{4.3.6b}$$

$$\Delta(180°) = \frac{24[\beta(G')^2 + \frac{1}{3}\beta(A)^2]}{c[45\alpha^2 + 7\beta(\alpha)^2]}, \tag{4.3.7}$$

where c is the light speed in vacuum. In the case of the right-angle (90°) geometry, the light scattered is polarized perpendicularly (Δ_x) or parallelly (Δ_z) to the scattering yz plane. The values present in the above expressions are polarizability tensor invariants obtained after averaging over all orientations of the molecule in the coordinate system. Specifically (4.3.8 – 4.3.9),

$$\alpha = \frac{1}{3}\alpha_{\alpha\alpha} = \frac{1}{3}(\alpha_{xx} + \alpha_{yy} + \alpha_{zz}), \tag{4.3.8}$$

$$G' = \frac{1}{3}G'_{\alpha\alpha} = \frac{1}{3}(G'_{xx} + G'_{yy} + G'_{zz}) \tag{4.3.9}$$

are isotropic invariants, and

$$\beta(\alpha)^2 = \frac{1}{2}\left(3\alpha_{\alpha\beta}\alpha_{\alpha\beta} - \alpha_{\alpha\alpha}\alpha_{\beta\beta}\right), \tag{4.3.10}$$

$$\beta(G')^2 = \frac{1}{2}\left(3\alpha_{\alpha\beta}G'_{\alpha\beta} - \alpha_{\alpha\alpha}G'_{\beta\beta}\right), \tag{4.3.11}$$

$$\beta(A)^2 = \frac{1}{2}\omega\alpha_{\alpha\beta}\varepsilon_{\alpha\gamma\delta}A_{\gamma\delta\beta} \tag{4.3.12}$$

are the anisotropic invariants (4.3.10 – 4.3.12).

The invariants are independent of the choice of origin, and each of them is an observable accessible to measurement. The $\varepsilon_{\alpha\gamma\delta}$ in (4.3.12) is the third-rank antisymmetric unit tensor. Since $\varepsilon_{\alpha\gamma\delta}A_{\gamma\delta\beta}$ is traceless, there is no corresponding isotropic tensor analogous to α and G'.

ROA measurements are carried out mostly in the $\theta = 180°$ geometry. This is due to the fact that when the molecule is composed entirely of idealized axially-symmetric bonds then $\beta(G')^2 = \beta(A)^2$ and $\alpha G' = 0$ and the ROA is generated exclusively by anisotropic scattering [9]. In this case the (4.3.5–4.3.7) expressions simplify to (4.3.13–4.3.15):

$$\Delta(0°) = 0, \tag{4.3.13}$$

$$\Delta_x(90°) = \frac{16\beta(G')^2}{c[45\alpha^2 + 7\beta(\alpha)^2]}, \tag{4.3.14a}$$

$$\Delta_z(90°) = \frac{4\beta(G')^2}{6c\beta(\alpha)^2}, \tag{4.3.14b}$$

$$\Delta(180°) = \frac{32\beta(G')^2}{c[45\alpha^2 + 7\beta(\alpha)^2]}. \tag{4.3.15}$$

Thus, one can see that the intensity of the ROA signal is the highest when the angle θ is 180°, and the smallest when $\theta = 0°$. This is significantly different than Raman measurements, where the intensity of Raman scattering is identical for both geometries. Because of weak ROA signals and a strong background, the ROA spectra of biological molecules in an aqueous medium are exclusively recorded at the $\theta = 180°$ geometry [8, 10]. It is possible to measure the ROA spectra for less demanding systems with different geometries, but this requires a rearrangement of the instrumental setup.

Quantum-chemical calculations of the ROA spectra are more demanding than the prediction of Raman spectra (*see* Chapter 7.11). In particular, special basis sets and methods should be used to correctly determine the magnetic parameters [11]. Moreover, one must also meticulously describe the effect of the solvent, which is particularly important for the interpretation of the spectra of samples dissolved in water [12]. Furthermore, to obtain reliable theoretical ROA spectra it is necessary to perform a detail conformational analysis.

4.3.3. Instrumentation

The three main laboratories where ROA spectrometers were developed in early ROA studies were located: in Glasgow (United Kingdom), led by Laurence Barron;

in Fribourg (Switzerland), led by Werner Hug; and at Syracuse University (USA), led by Laurence Nafie. In 1973 the first ICP-ROA bands were observed for α-phenylethanol, α-phenylethylamine, and α-phenylethylisocyanate [13, 14]. The first entire ROA spectrum was registered in 1975 by Hug et al. for (-)-α-pinene and (+)-α-phenylethylamine [15]. The first original measurements were performed with the ICP right angle depolarized scattering arrangement with the argon laser.

Almost a decade later, the first backscattering ICP spectrometer was built by Hug [4]. However, the spectrometer was damaged by a fire and the experiments were continued only after 1989 in Glasgow, using a new spectrometer equipped with the CCD (multi-channel charge-coupled device) detector [16]. Almost concurrently, the scattered circular polarization SCP-ROA spectrum was registered in Syracuse using a right-angle scattering geometry and in-plane linearly polarized incident radiation [17]. In these experiments, the ROA spectrum was measured as the difference of RCP and LCP Raman scattered radiation. However, they were measured sequentially and separately using the same state of linear polarization incident on the spectrograph and detector. In 1999, Hug and Hangartner proposed a method for measuring SCP-ROA with the simultaneous collection of the RCP and LCP scattered radiation [18]. This spectrometer, after the elimination of certain artifacts [19], became the prototype of the first commercially available instrument to measure the ROA spectra, launched in 2003 by BioTools Inc. The diagram of the latest ChiralRAMAN-2X™ spectrometer from BioTools Inc. is shown in Fig. 4.3.2.

The ROA instrument share a laser, a spectrograph, and a data acquisition system as in the ordinary Raman instrumentation [20]. It uses a 532 nm visible

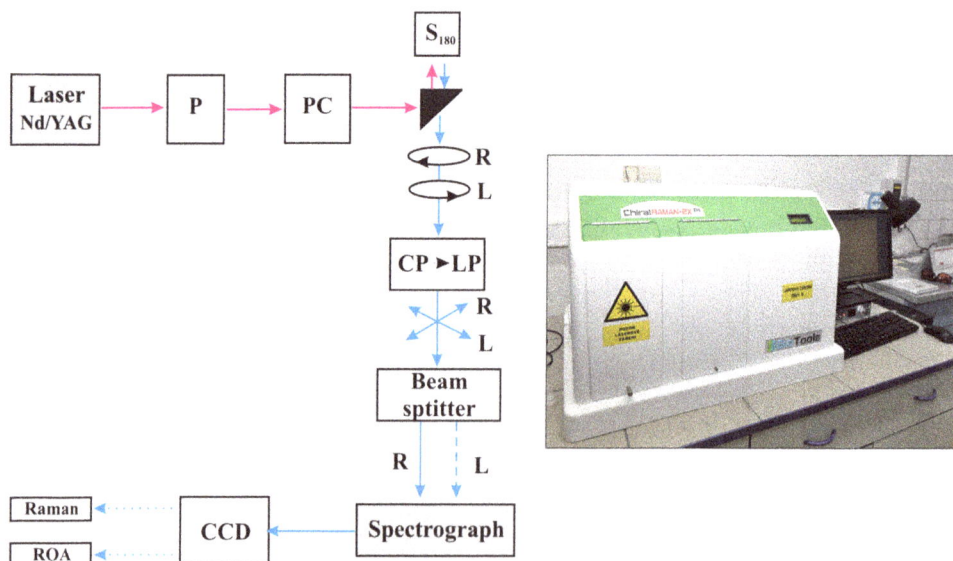

Fig. 4.3.2. The optical diagram of the ChiralRAMAN-2X™ instrument from BioTools Inc. The incident light is depicted in red while the scattered light in blue. (Reproduced from Ref. [3] with kind permission of John Wiley and Sons)

beam obtained from a frequency-doubled continuous Nd:YAG laser. The beam is first polarized (P) and then sent through a set of polarization conditioning (PC) optics to eliminate all linear polarization states and any residual circular polarization for the SCP-ROA measurement [3]. Thus, the beam that arrives at the sample is effectively unpolarized. Next, the incident laser radiation is directed to the sample in the backscattering geometry (S_{180}). The backscattering radiation passes through the prism and is directed to the scattered-radiation optical rail. The orthogonal (CP) circular polarization states of the scattered radiation are converted into orthogonal (LP) linear polarization states by a quarter-wave plate. The LP states are either selected by an analyzing polarizer for a separate measurement of the intensities for RCP and LCP radiation, or they are separated by a beam splitting cube for simultaneous measurements of RCP and LCP intensities [3]. They are displayed on the upper and lower halves of a CCD detector. Finally, the ROA spectrum is obtained by subtracting the RCP intensity from the LCP intensity, while the Raman spectrum is obtained by adding these two spectral accumulations [3].

The commercially available Chiral*RAMAN-2X*™ ROA spectrometer can be used for measurements of liquids in the 2000 – 100 cm^{-1} range of Raman shifts. However, several laboratories (e.g. in Prague, Glasgow and Tokyo) possess homemade spectrometers designed for specific applications. They differ in the ROA registration form, scattering geometry, and the frequency of the excitation beam.

4.3.4. Applications of ROA

Both the Vibrational Circular Dichroism and Raman Optical Activity techniques can be used to determine the absolute configuration (AC) of a chiral molecule [21]. In establishing absolute configuration, VCD and ROA are alternatives to X-ray crystallography. They permit for AC determination for a neat liquid or dissolved sample and require no pure single crystal to be obtained. However, because of a lack of empirical rules linking the sign and magnitude of the observed VCD or ROA intensities to the molecular structure, conformation, H-bonding, and solvation, the interpretation of the chiroptical spectra requires reliable quantum chemical calculations.

Moreover, due to the large number of bands, vibrational chiroptical techniques have an advantage over Electronic Circular Dichroism (ECD) – the other chiroptical method derived from electron spectroscopy UV-Vis [22]. The ECD spectra are composed of few bands which sometimes makes it difficult to unambiguously assign of the absolute configuration bands, particularly in the case of different diastereomers of the same compound. Furthermore, to register the UV-Vis and ECD spectra the chiral compounds must bear chromophore groups.

Nowadays, ROA is mainly used for the characterization of biomolecules such as peptides, proteins, carbohydrates and nucleic acids [23-26]. The ROA spectrum can provide not only information on the secondary structure such as α-helix or β-sheet [27, 28] but also on the tertiary fold [25, 29].

The ROA spectroscopy can also be applied for conformational studies [30, 31]. An example of such is nicotine, and (-)-nicotine, the sole enantiomer present in Nature. The pyridine ring and the methyl group can either be in *cisoid* or *transoid* position with respect to each other [31]. Moreover, the rotation about the bond connecting the two rings allows for additional modification, thus four different conformers can coexist in amounts which depend on the solvent [31]. The two neutral *transoid* conformers (Fig. 4.3.3) account for 99.9% of nicotine in the gas phase. It is difficult to experimentally evidence and distinguish them. However, the ROA spectrum of nicotine indicates that such an identification is possible [31]. The presence of the two individual *transoid* conformers can be confirmed in water by the presence of a doublet with maxima at 404 and 390 cm^{-1} (Fig. 4.3.3). Moreover, the band intensity ratio of ~1:1 suggests that in water the two conformers are present in approximately equal concentrations.

4.3.5. Summary

Raman optical activity (ROA) is used for structural studies of liquid chiral molecules. It measures the difference in intensity between Raman scattered right and left circularly polarized incident light. As ROA is the differential intensity spectrum,

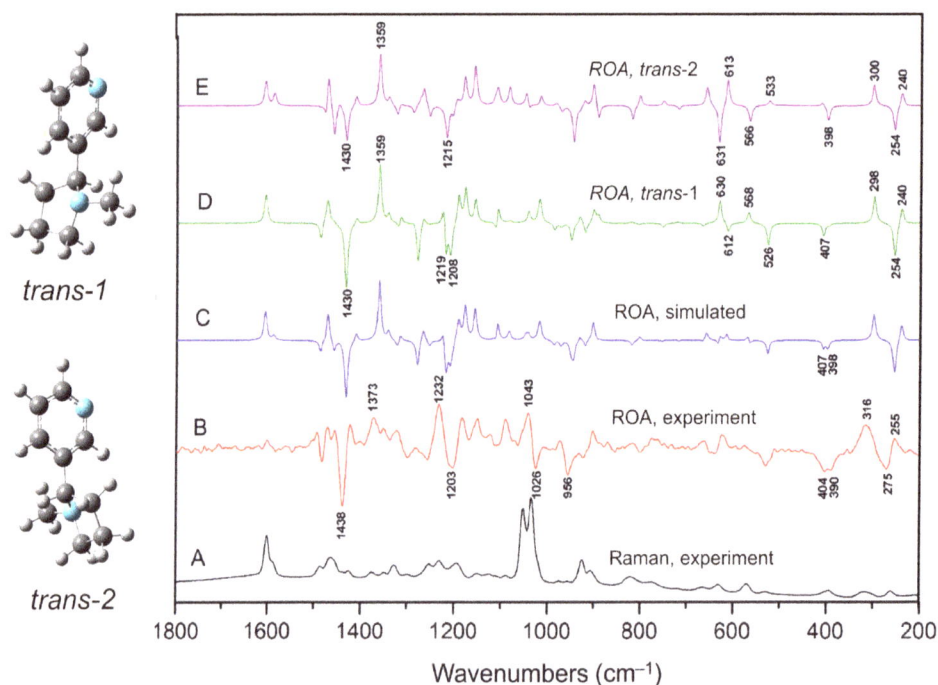

Fig. 4.3.3. The structures of the two most stable (-)-nicotine conformers and their Raman and ROA spectra. (Reproduced from Ref. [31] with kind permission of John Wiley and Sons)

its signals are low (10^{-3}-10^{-5}) compared to the parent Raman spectrum. Thus, the measurements require higher concentrations of the sample and a much longer accumulation time of the spectrum (sometimes a dozen or even several dozen hours).

There is a lack of empirical rules linking the sign and magnitude of the observed ROA intensities to the molecular structure, conformation, H-bonding, and solvation, therefore the interpretation of the chiroptical spectra requires a reliable quantum chemical calculations [3, 23].

ROA, like its parent Raman method, is a two-photon process. It creates opportunities for different types of ROA measurements connected with distinct polarizations of the incident and scattered beams. Also, as in the Raman experiments, there are three preferred scattering geometries: the right-angle scattering ($\theta=90°$), the forward scattering ($\theta=0°$), and the backward scattering ($\theta=180°$), where θ is the angle between the incident and scattered radiation. Chiral*RAMAN-2X*™, the only commercially available ROA spectrometer, can be used for measurements of liquids in the 2000 – 100 cm^{-1} range of Raman shifts with the backward scattering ($\theta=180°$). Moreover, it is equipped with the Nd/YAG laser with an excitation line at 532 nm.

Lately, much attention has been given to the ROA technique because of its important applications not only in the determination of the absolute configuration or structural/conformational studies, but even more so because of its potential in the study of biomolecules in an aqueous solution. An increase in the ROA signal intensities is expected in the currently developing Surface-Enhanced (SEROA) [32-34] and Resonance (RROA) [35-37] Raman Optical Activity techniques. The reproducibility of results and the registration of the SEROA and RROA spectra, which consist of the minor signals of enantiomers, is still a challenge. Yet there is no doubt that both of these techniques have considerable potential in the investigation of chiral samples.

References

1. Atkins P. W., Barron L. D., *Rayleigh scattering of polarized photons by molecules*, Mol. Phys., **16**, 453 (1969).
2. Barron L. D., Buckingham A. D., *Rayleigh and Raman scattering from optically active molecules*, Mol. Phys., **20**, 1111 (1971).
3. Nafie L. A., *Vibrational Optical Activity*, John Wiley & Sons Ltd., 2011.
4. Hug W., in: *Raman Spectroscopy* (ed. J. Lascombe), John Wiley & Sons, Ltd, Chichester 1982, pp. 3-12.
5. Nafie L. A., Freedman T. B., *Dual circular polarization Raman optical activity*, Chem. Phys. Lett., **154**, 260 (1989).
6. Yu G.-S., Nafie L. A., *Isolation of preresonance and out-of-phase dual circular polarization Raman optical activity*, Chem. Phys. Lett., **222**, 403 (1994).
7. Li H., Nafie L.A. *Simultaneous acquisition of all four forms of circular polarization Raman optical activity: results for a-pinene and lysozyme*, J. Raman Spectrosc., **43**, 89 (2012).
8. Barron L. D., Buckingham A. D., *Vibrational optical activity*, Chem. Phys. Lett., **492**, 199 (2010).

9. Barron L. D., Vrbancich J., in: *Topics in Current Chemistry*, Springer-Verlag, vol. 123, 1984, pp. 151-182.

10. Hecht L., Barron L. D., Hug W., *Vibrational Raman optical activity in backscattering*, Chem. Phys. Lett., **158**, 341 (1989).

11. Cheeseman J. R., Frisch M. J., *Basis Set Dependence of Vibrational Raman and Raman Optical Activity Intensities*, J. Chem. Theory Comput., **7**, 3323 (2011).

12. Hopmann K. H., Ruud K., Pecul M., Dračínský M., Bouř P., *Explicit Versus Implicit Solvent Modeling of Raman Optical Activity Spectra*, J. Phys. Chem. B, **115**, 4128 (2011).

13. Barron L. D., Bogaard M. P., Buckingham A. D., *Raman scattering of circularly polarized light by optically active molecules*, J. Am. Chem. Soc. **95**, 603 (1973).

14. Barron L. D., Bogaard M. P., Buckingham A. D., *Differential Raman Scattering of Right and Left Circularly Polarized Light by Asymmetric Molecules*, Nature, **241**, 113 (1973).

15. Hug W., Kint S., Bailey G. F., Scherer J. R., *Raman circular intensity differential spectroscopy. Spectra of (-)-.alpha.-pinene and (+)-.alpha.-phenylethylamine*, J. Am. Chem. Soc. **97**, 5589 (1975).

16. Barron L. D., Hecht L., Hug W., MacIntosh M. J., *Backscattered Raman optical activity with a CCD detector*, J. Am. Chem. Soc., **111**, 8731 (1989).

17. Spencer K. M., Freedman T. B., Nafie L. A., *Scattered circular polarization Raman optical activity*, Chem. Phys. Lett., **149**, 367 (1988).

18. Hug W., Hangartner G., *A novel high-throughput Raman spectrometer for polarization difference measurements*, J. Raman Spectr., **30**, 841 (1999).

19. Hug W., *Virtual enantiomers as the solution of optical activity's deterministic offset problem*, Appl Spectrosc., **57**, 1 (2003).

20. Hug W., in: *Comprehensive Chiroptical Spectroscopy, vol. 1, Instrumentation, Methodologies, and Theoretical Simulations*. (eds. N. Berova, P. L. Polavarapu, K. Nakanishi, R. W. Woody), J. Wileys & Sons, 2012, pp. 147-177.

21. He Y., Wang B., Dukor R. K., Nafie L. A., *Determination of Absolute Configuration of Chiral Molecules Using Vibrational Optical Activity: A Review*, Appl Spectrosc., **65**, 699 (2011).

22. Berova N., Di Bari L., Pescitelli G., *Application of Electronic Circular Dichroism in Configurational and Conformational Analysis of Organic Compounds*, Chem. Soc. Rev., **36**, 914 (2007).

23. Chruszcz-Lipska K., Blanch E. W., in: *Optical Spectroscopy and Computational Methods in Biology and Medicine (ed. M. Barańska)*, Springer, vol. 14, 2013, pp. 61-81.

24. Barron L. D., Hecht L., in: *Comprehensive Chiroptical Spectroscopy, vol. 2, Applications in Stereochemical Analysis of Synthetic Compounds, Natural Products, and Biomolecules*, (eds. N. Berova, P. L. Polavarapu, K. Nakanishi, R. W. Woody), J. Wileys & Sons, 2012, pp. 759-793.

25. Yamamoto S., Kaminský J., Bouř P., *Structure and vibrational motion of insulin from Raman optical activity spectra*, Anal. Chem., **84**, 2440 (2012).

26. Unno M., Kikukawa T., Kumauchi M., Kamo N., *Exploring the Active Site Structure of a Photoreceptor Protein by Raman Optical Activity*, J. Phys. Chem. B, **117**, 1321 (2013).

27. Weymuth T., Reiher M., *Characteristic Raman optical activity signatures of protein β-sheets*, J. Chem. Phys. B, **117**, 11943 (2013).

28. Barron L. D., Hecht L., McColl I. H., Blanch E. W., *Raman Optical Activity Comes of Age*, Mol. Phys., **102**, 731 (2004).

29. Blanch E.W., Hecht L., Syme C. D., Volpetti V., Lomonossoff G. P., Nielsen K., Barron L. D., *Molecular structures of viruses from Raman optical activity*, J. Gen. Virol., **83**, 2593 (2002).

30. Nieto-Ortega B., Casado J., Blanch E. W., López Navarrete J. T., Quesada A. R., Ramírez F. J., *Raman Optical Activity Spectra and Conformational Elucidation of Chiral Drugs. The Case of the Antiangiogenic Aeroplysinin-1*, J. Phys. Chem. A, **115**, 2752 (2011).

31. Barańska M., Dobrowolski J. Cz., Kaczor A., Chruszcz-Lipska K., Gorza K., Ryguła A., *Tobacco alkaloids analyzed by Raman spectroscopy and DFT calculations*, J. Raman. Spectrosc., **43**, 1065 (2012).
32. Abdali S., Blanch E. W., *Surface enhanced Raman optical activity (SEROA)*, Chem. Soc. Rev., **37**, 980 (2008).
33. Novák V., Šebestík J., Bouř P., *Theoretical Modeling of the Surface-Enhanced Raman Optical Activity*, J. Chem. Theory Comput., **8**, 1714 (2012).
34. Chulhai D. V., Jensen L., *Simulating Surface-Enhanced Raman Optical Activity Using Atomistic Electrodynamics-Quantum Mechanical Models*, J. Phys. Chem. A, **118**, 9069 (2014).
35. Luber S., Neugebauer J., Reiher M., *Enhancement and de-enhancement effects in vibrational resonance Raman optical activity*, J. Chem. Phys., **132**, 044113 (2010).
36. Merten C., Li H., Nafie L. A., *Simultaneous Resonance Raman Optical Activity Involving Two Electronic States*, J. Phys. Chem. A, **116**, 7329 (2012).
37. Zając G., Kaczor A., Chruszcz-Lipska K., Dobrowolski J. Cz., Barańska M., *Bisignate resonance Raman optical activity: a pseudo breakdown of the single electronic state model of RROA?*, J. Raman Spectrosc., **45**, 859 (2014).

4.4. Raman imaging

Agnieszka Kaczor

In the 1970s and 1980s of the XX century considerable development of the charge-coupled devices (CCDs) was achieved [1,2]. Due to their high quantum efficiency, application of CCD cameras in Raman spectroscopy have brought the significant reduction of the measurement time compared to experiments with photomultipliers as signal detectors [3]. This technical progress was the very reason of development of special techniques of Raman spectroscopy aimed at not only identification of compounds in the sample, but also analysis of their distribution in space.

The original concept of Raman imaging* is rather simple. Registration of Raman spectra in a given (studied) volume of the sample is followed by the analysis of the marker bands for a given compound (compounds) in the obtained dataset of Raman spectra. Such analysis is based on calculations of the integral intensity of marker bands that results in determination of the distribution of compounds related to these marker bands in the studied volume of the sample (Fig. 4.4.1). Raman imaging gives simultaneously information about multiple components in the sample unless their marker bands overlap. Additionally, the method is non-destructive and does not require sample preparation, therefore it is a convenient technique to *in situ* studies[4,5]. Raman imaging is generally preformed in the setup in which the sample is shifted "point by point" (point imaging) in respect to the objective. If a microscope is not used in the setup, such technique can be called imaging in the macro scale, contrarily to imaging *via* a microscope (also known as "Raman microscopy"), described below.

---

* Nowadays terms "mapping" and "imaging" are used interchangeably, due to the fact that the final result in both cases is the image (of components' distribution in the sample). Classically, the distinction between these two terms was based on the illumination technique. The term "mapping" was reserved for the "point by point" illumination and registration of the spectra, while for linear or general illumination the term "imaging" was used.

First Raman spectrometers coupled with a microscope appeared in the 1990s of the XX century. The microscope is used to transfer both the incident (laser) light and the Raman scattered light, therefore the obtained information is derived from a very small volume of the sample called voxel (*volumetric element*, schematically illustrated in Fig. 4.4.1A as a purple ellipsoid) that due to illumination is a source of scattered light (Fig. 4.4.1 B). Several thousands of spectra are finally registered in a single imaging experiment.

Exactly speaking, the spatial resolution depends both on the excitation wavelength of the laser (λ) and the numerical aperture* (*NA*) of the microscope objective.

The lateral resolution (Δx) is the smallest distance between two distinguishable points, given by the expression (4.4.1):

$$\Delta x = 0{,}61 \, \frac{\lambda}{NA}. \tag{4.4.1}$$

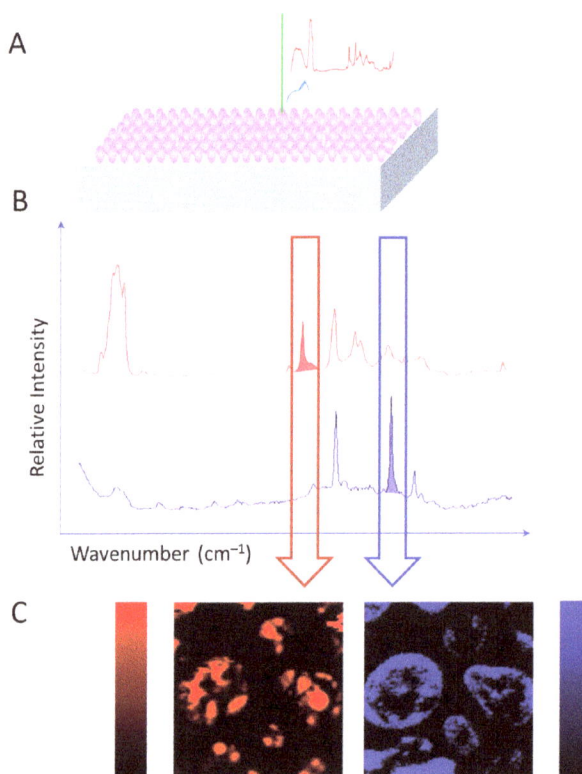

Fig. 4.4.1. Registration of the Raman spectra set in the given volume of the sample (A, top layer of the sample, denoted by purple color) provides information about distribution of the sample components in the studied fragment. Analysis of the marker bands (in this case bands of two different components denoted in different colours, (B)) in all the registered spectra results in determination of components' distribution presented in the form of color-coded images (C)

---

* Numerical aperture (*NA*) is a number characterizing an objective and defined by the equation,: $NA = n \sin\theta$, where n is the index of refraction of the medium in-between the sample and the objective, and θ is the angle between the marginal ray, entering the objective and the optical axis of its lens.

This equation indicates that to minimize a lateral resolution (to achieve a high resolving power) it is necessary to apply objectives of a high numerical aperture and light sources in the low-wavenumber range.

The volume of the voxel is also defined by the depth resolution. Confocal microscopes are usually applied in the Raman imaging in order to achieve a good depth resolution. In confocal microscopy, the signal from above or below the focal plane does not contribute to the image. The basic principle of a confocal microscope is illustrated below (Fig. 4.4.2).

As the result of the application of a small pinhole before a detector (in practice frequently accomplished by application of a fiber of a small dimension, typically 10–100 mm), the detected light is restricted to a small volume around the focal point and the radiation scattered from the planes other than the focal plane is eliminated. Although the depth resolution (Δz), similarly to the lateral resolution, depends on the wavelength of the incident light and the numerical aperture of the objective, the dependence has a different form (4.4.2):

$$\Delta z = \frac{\lambda}{NA^2}. \tag{4.4.2}$$

Only point imaging is an option enabling confocal measurements. As it was mentioned, in this technique a point by point registration of the signal is achieved by moving the scan table in respect to the exciting laser. Scanning movement is controlled by the computer, and the resolution of the table in the case of piezo--scanners can be as even less than 1 nm. The total measurement time is the sum of the illumination time of the sample and the time necessary for the movement

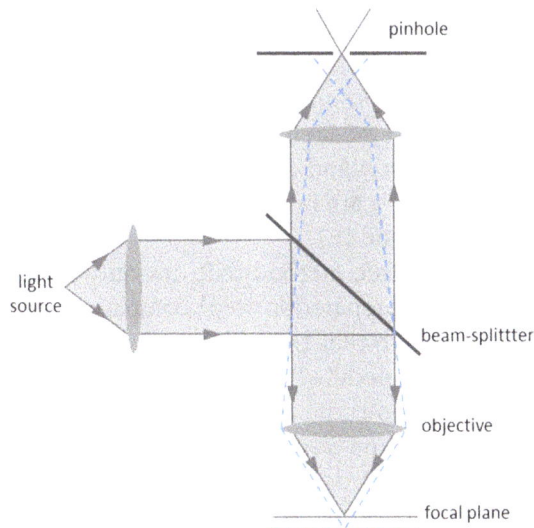

Fig. 4.4.2. The basic principle of a confocal microscope. The light emitted by the point light source is focused on a beam-splitter and, after being reflected, is directed to the sample. A scattered light goes back *via* an objective and the beam-splitter and hits the aperture with a small pinhole. Only the light from the focal plane is transmitted to a detector (the light from other planes is eliminated)

of the table. The measurement time can be reduced in the linear or global illumination modes. In the linear illumination a complete line on the sample is illuminated and projected onto the entrance slit of the spectrometer. The vertical axis of the CCD camera is used for position information and the horizontal axis for scattering intensity and in this mode hundreds of spectra can be registered at the same time [5]. In the global illumination all the measured area of sample is illuminated at once and the signal is projected onto an imaging CCD. In such a setup, the whole range of Raman spectrum cannot be registered at the same time, therefore, the spectral information is obtained by recording several images in different wavenumber ranges [5].

In the confocal imaging Raman experiment, hundreds of thousands of Raman spectra are routinely recorded in a single experiment. Every spectrum contains the whole information about the content and structure of all components in the studied voxel that can be very difficult to extract in the case of complex samples. Therefore, univariate methods such as a simple analysis of the marker bands (Fig. 4.4.1C), may not be effective. Therefore, in data analysis both careful pre-processing (baseline correction, cosmic spikes removal, smoothing) followed by chemometric methods of multivariate analysis, particularly, cluster analysis, are often used. Methods of data analysis are described in detail in Chapter 5.

Applications of Raman imaging are very broad. This technique is successfully applied in mineralogy, petrography, to study artworks, pharmaceuticals and polymers. More and more commonly Raman imaging is used in biological sciences. As the lateral resolution (even below 0.3 mm in the case of 488 nm excitation) of Raman microscopy is compared with the size of subcellular structures, this method is well suited to study cell components and biochemical processes occurring in cells and tissues [6-11]. For example, Fig. 4.4.3 shows Raman images showing distribution of chosen biocomponents in the cross section of the murine aorta (A) compared with the images of the immunohistochemically stained tissue (B) (a gold standard) [6]. A single measurement enables determination of the distribution of, among others, cell nuclei, elastin and organic components of the sample and the obtained Raman images (C,D,F) are in the very good agreement with the fluorescence image of the stained tissue (E).

Raman imaging is a label-free method that not only reduces the time of the analysis (eliminates a sample preparation step) compared to immunohistochemical staining, but also guarantee that the information is derived from chemically non-modified (or only slightly modified) material. Moreover, obtained Raman spectra contain the information about the molecular structure of the sample components, which is not available in the case of usage of histo(patho)logical techniques.

Application of a confocal microscope coupled with the Raman spectrometer and depth scanning of the successive thin (about 1 mm) layers of the sample enables reconstruction of its 3D image [7-9]. The example of 3D imaging, showing the reconstruction of the lipid structure in the innermost layer of the murine aorta wall in given in Fig. 4.4.4 [9].

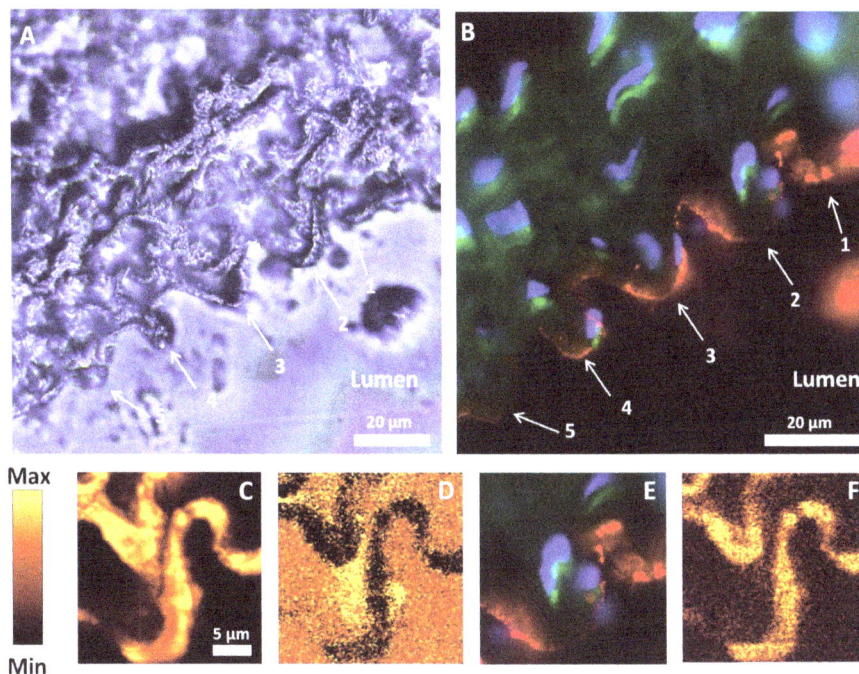

Fig. 4.4.3. The visible tissue image (40×, A) along with the fluorescence image of the immuno-histochemically stained tissue (40×, B, E): red, blue and green denote endothelium, cell nuclei and elastin, respectively) of the fragment of the murine aorta cross-section; Raman images showing distribution of all organic components (C), cell nuclei (D) and elastin (F) in the area respective to the area in E. Reprinted from *Biomed. Spectrosc. Imag.*, 2, M. Pilarczyk, A. Rygula, L. Mateuszuk, S. Chlopicki, M. Baranska, A. Kaczor, *Multi-methodological insight into the vessel wall cross-section: Raman and AFM imaging combined with immunohistochemical staining*, **2**, 191 (2013), with permission from IOS Press, after modification.

Raman imaging (also in the 3D variant) has various potential applications in the field of medical diagnostics. The analysis of the microscopic fragments of the cells or tissues with the good spatial resolution is used to study biochemical changes in various pathologies, among others tumours, atherosclerosis, diabetes and aortic valve stenosis [8-11]. Such studies give insight into molecular mechanisms governing these diseases, also at the early stage of their development. Intraoperative determination of tumour boundaries is another promising application of Raman imaging [12].

The wide range of benefits of Raman microscopy (high spatial resolving power, not destructive, label-free, versatile, of high chemical and structural specificity) makes it a method of big prospects and multiple applications both in science and industry.

Examples of application of Raman imaging are given in Chapters 7.3, 7.13-7.15 and univariate and multivariate methods of data analysis are presented in Chapter 5.

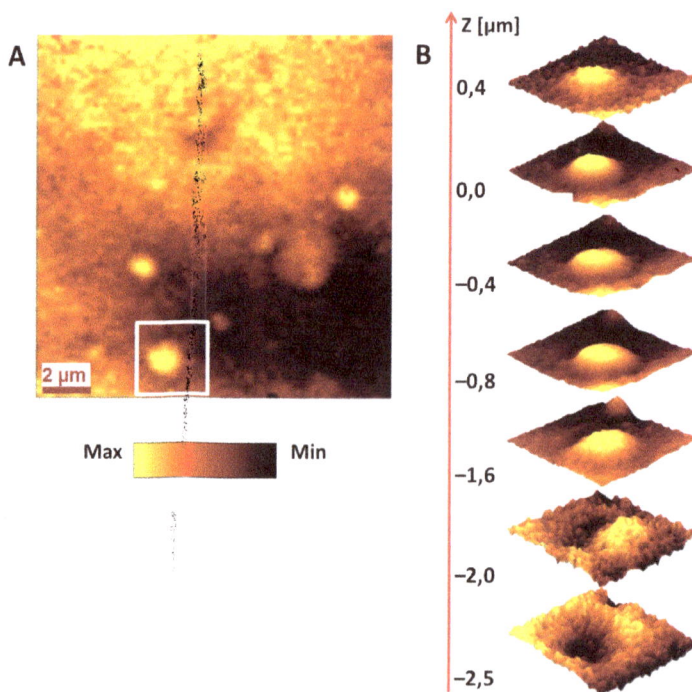

Fig. 4.4.4. The Raman image showing distribution of lipids in the fragment of the innermost murine aorta wall (db/db diabetes model, (A)) and the result of the depth imaging of the chosen fragment of the sample (the with rectangular in A) enabling estimation of the volume of the studied lipid structure and reconstruction of its 3D image (B); based on [8] *Creative Commons Attribution (CC BY)* license

References

1. Press. W. H., Teukolsky S. A., Vetterling W. T., Flannery B. P. *Savitzky-Golay Smoothing Filters* in *Numerical Recipes in C*, The Press Syndicate of the University of Cambridge, UK 1999, pp. 650-655.
2. Ramos P., Ruisánches I., *Noise and background removal in Raman spectra of ancient pigments using wavelet transform*, J. Raman Spectrosc., **36**, 848 (2005).
3. Hollricher O., in: *Confocal Raman Microscopy* (eds. T. Dieing, O. Hollricher, J. Toporski) Springer-Verlag, Germany 2010, pp. 44-45.
4. Kaczor A., Turnau K., Barańska M., *In situ Raman imaging of astaxanthin in a single micro-algal cell*, Analyst, **136**, 1109 (2011).
5. Barańska M., Baranski R., Grzebelus E., Roman M., *In situ detection of a single carotenoid crystal in a plant cell using Raman microspectroscopy*, Vib. Spectrosc., **56**, 166 (2011).
6. Pilarczyk M., Ryguła A., Mateuszuk L., Chlopicki S., Barańska M., Kaczor A., *Multi-methodological insight into the vessel wall cross-section: Raman and AFM imaging combined with immunohistochemical staining*, Biomed. Spectrosc. Imag., **2**, 191 (2013).
7. Majzner K., Kaczor A., Barańska M., *3D confocal Raman imaging of endothelial cells and vascular wall: perspectives in analytical spectroscopy of biomedical research*, Analyst, **138**, 603 (2013).

8. Pilarczyk M., Czamara K., Barańska M., Natorska J., Kapusta P., Undas A., Kaczor A., *Calcification of aortic human valves studied in situ by Raman microimaging: following mineralization from small grains to big deposits*, J. Raman Spectrosc., **44**, 1222 (2013).

9. Pilarczyk M., Mateuszuk L., Ryguła A., Kepczynski M., Chlopicki S., Barańska M., A. Kaczor A., *Endothelium in Spots – High-Content Imaging of Lipid Rafts Clusters in db/db Mice*, PLOS One, **9**, e106065 (2014).

10. Rygula, A. Pacia M. Z., Mateuszuk L., Kaczor A., Kostogrys R., Chlopicki S., Barańska M., *Identification of a biochemical marker for endothelial dysfunction using Raman spectroscopy*, Analyst, **140**, 2185 (2015).

11. Abramczyk H., Brozek-Pluska B., *Raman Imaging in Biochemical and Biomedical Applications. Diagnosis and Treatment of Breast Cancer*, Chem. Rev., **113**, 5766 (2013).

12. Old O. J., Fullwood L. M., Scott R., Lloyd G. R., Almond L. M., Shepherd N. A., Stone N., Barr H., Kendall C., *Vibrational spectroscopy for cancer diagnostics*, Anal. Meth., **6**, 3901 (2014).

5

Chemometric analysis of FT-IR and Raman spectra

Katarzyna Majzner, Kamila Kochan, Małgorzata Barańska

Raman and FT-IR imaging (both described in Chapters 3.1 and 4.4, respectively) lead to obtaining large amounts of spectral and spatial data. For the analysis two different approaches are used: (1) classic, based on the integration of selected marker bands and (2) chemometric based on classification, such as cluster analysis.

Classical approach, which is the analysis of chemical composition based on the marker bands present in the spectrum, allows extracting only part of the information about the complex systems, to which biological samples undoubtedly belong. The basic requirement for this type of analysis is the knowledge of the position of the analyte marker bands and their presence in the spectra along with the possibility of their identification. Chemometric methods, on the other hand, are in general multi-dimensional analysis techniques, combined with grouping of pixels, which is based on the similarity of their spectral properties. This approach facilitates the analysis of the results, when the analyte marker bands are not clearly visible and/or it is not *a priori* known what chemical changes should be expected in the recorded spectra.

The method of cluster analysis in relation to the classical approach, and its application on the example of the spectra obtained by Raman imaging of a single cell is presented below.

5.1. Analysis of marker bands

Classical analysis of the results obtained by means of a imaging relies on calculating the integral value (area under the band) in the ranges of the selected bands appearing in the spectra. The results of this analysis are the colour drawings on a scale where a value of the integral is color-coded according to the selected scale. As a result, a three-dimensional matrix is obtained (Figure 5.1), where x and y indicate the location of each pixel on the image and the z – dimension contains information about of the integral value of the band at that point (keep in mind that each point on the image is represented as a spectrum).

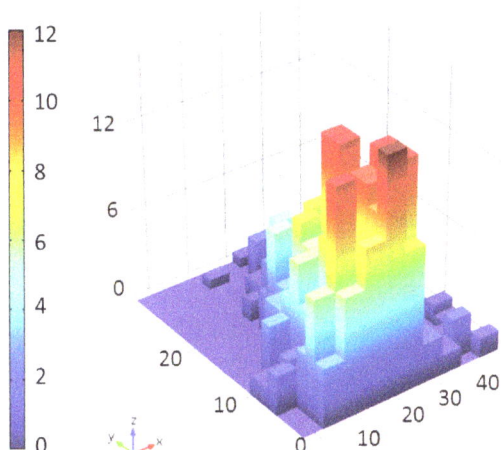

Fig. 5.1. Schematic diagram of the analysis based on the integration of the band characteristic for certain substance

The purpose of the analysis is to identify areas of the sample, for which the corresponding spectra are characterized by high intensity of the analysed bands. Figure 5.2 shows the results of such analysis of spectra obtained from Raman imaging of a single cell for the two bands: related to the C-H stretching vibration, ν_{C-H} (Fig. 5.2A & 5.2B) and the stretching vibration of the O-P-O, ν_{O-P-O} (Fig. 5.2C & 5.2D).

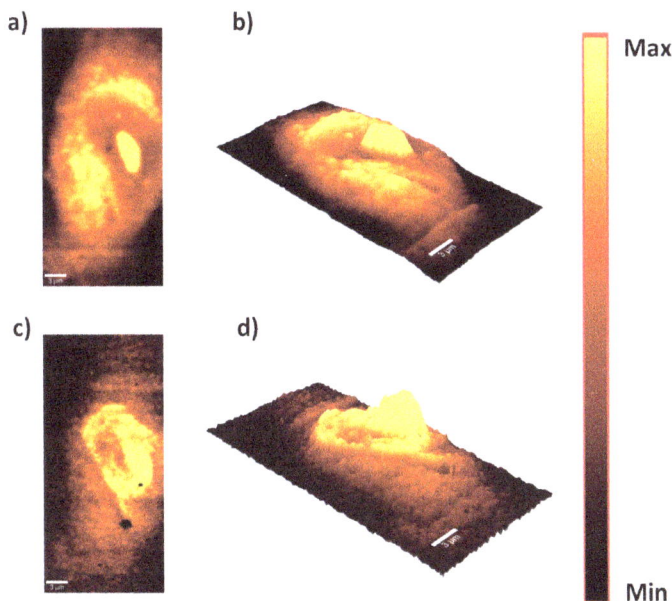

Fig. 5.2. Distribution of selected components of a cell, constructed on the basis of the integrated intensity of Raman bands associated with the C-H stretching vibrations in the range of 2800–3020 cm^{-1} (A, B) and O-P-O stretching vibrations in the range of 1080-1118 cm^{-1} (C, D). The measured cell is an endothelial cell from an EA.hy926 cell line. Measurement parameters: 532 nm laser line, the integration time of 0.5 s, the size of the imaged area of 16.7 × 34.1 µm², the spatial resolution of 0.32 microns (image size: 52 × 106 points). Scale bar is presented on the right side

The results of the presented analysis are images coded in yellow-brown colour scale, where the bright colour (yellow) corresponds to a high value of integrated Raman intensity and dark (brown and black) – to a low value. Successively integrating Raman bands of interest it is possible to follow the distribution of the major components of the cell. The range 2800–3020 cm^{-1} allows for visualization of the entire imaged cell, as it includes vibrations deriving from proteins and lipids. Additionally, the intensity of these bands is related to the thickness and density of the sample. Spectral ranges 785–810 and 1080–1118 cm^{-1} provide information about the distribution of nucleic acids in the cell and therefore indicate the shape, size and position of the nucleus (the signal from the nucleic acid may also be localized in the cytoplasm, where it comes from the ribosomal, matrix and transfer RNA).

5.2. Cluster analysis (CA)

Cluster Analysis (CA) is a method of grouping objects similar to each other. In the case of the use of cluster analysis as a method of data mining for data recorded using Raman and FT-IR imaging, an object is understood as a single spectrum assigned on the image to a single pixel. Consequently, as a result of CA a number of groups (referred to as "classes" or "clusters") are obtained. Every class is in fact a group of spectra with similar spectral profile. Objects/spectra within the same group should be maximally similar and uniform in characteristics defining this particular cluster (the profile of the spectrum in the selected spectral range) and maximally different from spectra belonging to other classes. The division into classes is based on a defined mathematical algorithm.

Each pixel of the image corresponds to a spectrum. Therefore, grouping of the spectra automatically results in grouping of pixels. The pixels of the same class are encoded in the same colour. Additionally, spectra of each class are averaged and presented as a single spectrum (coded with the same colour as the colour of the pixels belonging to this class).

Different algorithms and cluster analysis methods can lead to different results of clustering. The classes can differ from each other e.g. in their size and distribution, which can be directly reflected not only in the resulting image, but also in the averaged spectra. Furthermore, the use of different clustering method allows identifying a new and important class, which in the case of cells may correspond e.g. to subcellular structure.

Cluster analysis methods can be divided according to several criteria. Each of them takes into account the difference between the characteristics of different groups generated by different algorithms. The basic classification includes hierarchical and non-hierarchical methods (Table 5.1) and will be discussed in the following sections.

Table 5.1. Classification of cluster analysis methods (names and abbreviations of methods will be discussed in the following sections)

Cluster Analysis	
Hierarchical	**Non -hierarchical**
Method: agglomerative, divisive	Method: variation, iterative
Example: HCA (agglomerative method)	Example: KMCA (variation method) FCA (iterative method)
Methods for calculating the distance: Manhattan, Euclidean, Pearson, Mahalonobis	
Clustering algorithms (linkage criteria): Single-linkage („nearest neighbour'), complete-linkage („furthest neighbour"), mean (or average) linkage, centroid linkage, Ward's method (Ward's minimum variance method)	

5.2.1. Hierarchical cluster analysis (HCA)

Hierarchical Cluster Analysis (HCA) is one of the most popular methods of clustering spectra. Hierarchical methods are based on creation of a hierarchy of classification. Dependently on the manner of classification two types of hierarchical methods can be distinguished:

a) agglomerative method,
b) divisive method.

The agglomerative clustering method begins with the treatment of each object as a separate class (*n* single-element classes, represented by A-F in Fig. 5.3). In subsequent steps of grouping similar objects are combined into clusters as follows: the distances between every two objects are calculated according to the chosen method for calculating the distance (see text below) and a pair/group of classes, between which the distance is the smallest (determined on the basis of chosen

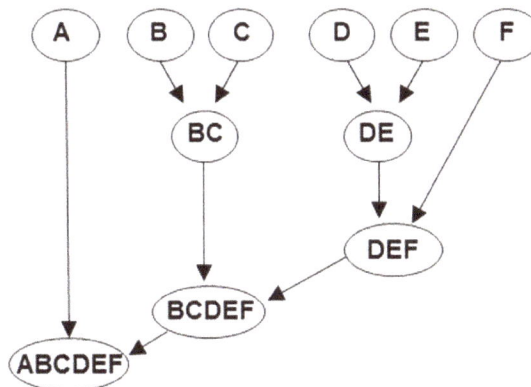

Fig. 5.3. Illustration of data classification by agglomerative method. If the arrows would be the other way the shame would represent a classification by divisive approach

linkage criterion) is found and combined in one. In the subsequent stages the procedure is repeated, resulting every time in reduction of the number of classes, until
all objects are combined into one common class. [1]

The divisive method starts from placing all objects in one class (single n-element
class). In the subsequent steps this class is divided into a number of sub-classes
until each object will be located in a separate class (n single-element classes). The
division is made on the basis of the highest value of dissimilarity (or the smallest
value of similarity).

Hierarchical cluster analysis results in obtaining a dendrogram showing the "origin" and the possibility of further separation of certain classes. The difference between agglomerative and divisive method lies primarily in the directionality of algorithms actions, while constructing a dendrogram. The end point of agglomerative
procedure corresponds to the start point in divisive method.

The distance between the objects ($d_n(x,y)$) (e.g. belonging to the same class) can
be calculated using several methods:

[1] Manhattan distance (also known as „city block distance")

$$d_n(\bar{x},\bar{y}) = \sum_{i=1}^{n} |y_i - x_i|,$$ (5.1)

[2] Euclidean distance

$$d_n(\bar{x},\bar{y}) = \sqrt{\sum_{i=1}^{n} (y_i - x_i)^2},$$ (5.2)

[3] Pearson correlation coefficient

$$\rho_{x,y} = \frac{\operatorname{cov}(x,y)}{\sigma_X \sigma_Y},$$ (5.3)

[4] Mahalanobis distance

$$d_n(\bar{x},\bar{y}) = \sqrt{\frac{\sum_{i=1}^{n} (y_i - x_i)^2}{\sigma_X \sigma_Y}},$$ (5.4)

where: x, y – are the representations of objects as points in the n-dimensional space,
σ – is the standard deviation between the sets of objects X and Y, cov – covariance.

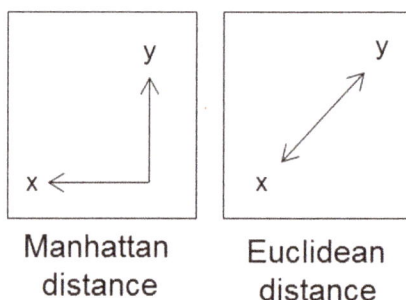

Fig. 5.4. Diagram showing the method
of calculating the distance in relation to the city
block (Manhattan) and the Euclidean metrics

Manhattan
distance

Euclidean
distance

There are many types of hierarchical clustering algorithms and among the most popular of the linkage criteria for calculating the similarity between objects (e.g. between: object – object, object – class) are:

a) single- linkage ("nearest neighbour") method, where the distance is defined as the smallest distance between objects belonging to two different groups,
b) complete-linkage ("furthest neighbour ") method, where the distance between the two classes is determined as the distance between the two most further located objects,
c) mean (or average) linkage method (occurring also as a median method), where the distance is measured as an average value of distances between all objects belonging to the two groups (and the average value can be calculated as mean or median),
d) centroid (center of gravity) method, which involves the determination of the geometric center of gravity (centroid) and calculating the distance between the two groups as the distance between their centroids,
e) Ward's method (Ward's minimum variance method), in which the estimation of the distance between the clusters is based on a variance approach (determination of the sum of squares of deviations within all groups and linking the groups for which the value is the smallest after linkage).

The agglomerative hierarchical cluster analysis is conducted according to the following procedure:

[1] Selection of a method for calculating the distance between the objects and the parameters describing the data set; calculation of the distance between all objects.
[2] Determination of the linkage (similarity) criterion and calculation of the similarity measure between all objects.
[3] Selection of the two of the most „similar" to each other objects and binds them in one class.
[4] Recalculation of the similarity measure of all the objects relative to the newly created class and each other.
[5] Repetition of the process until the all objects are assigned to one class – the final result is a dendrogram (an example of which is presented in Fig. 5.5).

Hierarchical methods, although they are extremely useful in the analysis of the Raman and FT-IR images, have their disadvantages and limitations. Primarily, they do not allow for correction of the already formed clusters. Therefore any wrong class assignment cannot be corrected in the next step of the algorithms' calculations. Furthermore, because of the number of steps performed during the analysis and the need for a full recalculation of distance and similarity at every step this type of analysis can be very time-consuming (as compared to the described below *k-means* analysis) and requires more computer memory.

An example of a Hierarchical Cluster Analysis of a Raman image of a single cell is shown in Figure 5.6. In the presented example the Manhattan (city block)

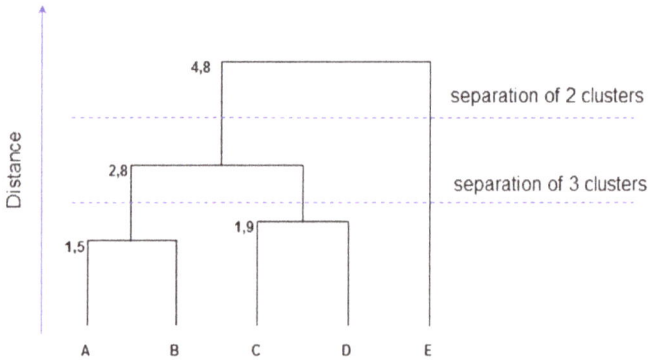

FIG. 5.5. An example of a dendrogram created as a result of HCA for the hypothetical a data set consisting of 5 elements, (e.g. 5 spectra) (A–E). Calculated distances (1.5, 1.9, 2.8 and 4.8) between clusters or objects grow from the bottom to the top. A person who analyses the data determines whether objects are divided into 2 or 3 classes

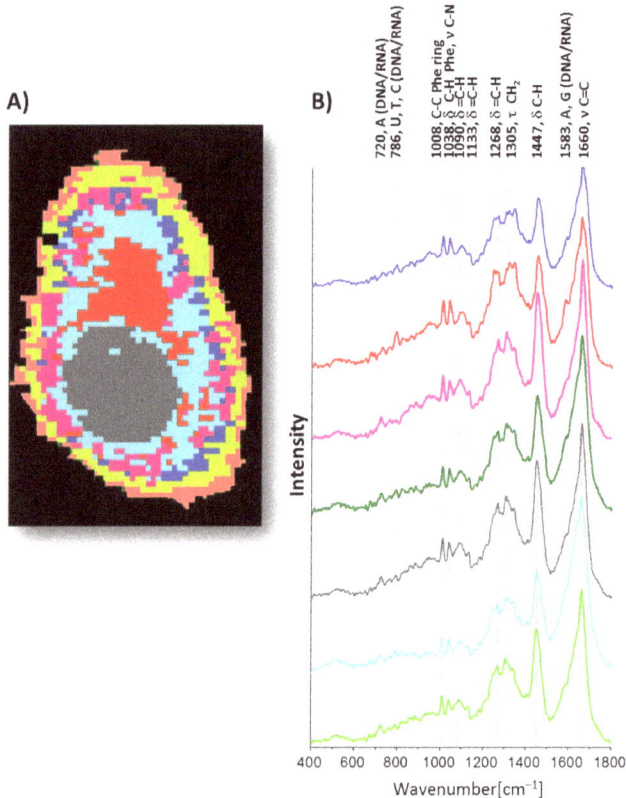

FIG. 5.6. Map of distribution of classes (A) along with their average spectra (B) obtained as a result of the hierarchical cluster analysis (HCA) performed with the use of an agglomerative algorithm with Ward's method (clustering algorithm) and Manhattan distance (method for calculating the distance). The measured cell is an endothelial cell from a line HAoEC. Measurement parameters: 488 nm laser line, the integration time 0.65s, size of the imaged area $38 \times 65 \ \mu m^2$, the spatial resolution: 0.8 microns (image size: 48×81 points)

method for calculating the distance and the Ward's algorithm for calculating the linkage criterion (clustering algorithm) were chosen.

5.2.2. Non-hierarchical cluster analysis

In the case of non-hierarchical cluster analysis the objects are grouped in a pre-established number of classes, without considering their hierarchy.

K-Means Cluster Analysis method (KMCA) is a very popular algorithm of cluster analysis. In contrast to HCA, it is based not on the measured distances themselves, but on the interclass variances between them. During the entire clustering process, the algorithm creates classes that are represented by their geometric means (centroids), so therefore by single points. This allows the use of a simple distance measures for calculating the distance and, at the same time, reduces significantly the amount of mathematical operations during every cycle (instead of calculating the distance between every two objects, only the distances between every object and every centroid is calculated). However, it requires to define a desired number of classes before their generation (this is also one of the disadvantages of this method).

The class division in the KMCA method starts from creating a prototype classes in the number initially specified in the input parameters (by the person performing the analysis). Each prototype class is created by the location of the centroid of class (sometimes random) and then by assigning the objects/spectra. Spectra from the analysed database are pre-allocated between the groups in accordance with the distance from the centroid (the spectrum goes to this class, whose centroid is the nearest). After generating prototype clusters, for each class a corrected centroid position is calculated. In the next step the objects are assigned again to classes according to their distance from the centroid. In this way the objects (spectra) belonging to the class are verified at each stage of grouping. This procedure is repeated until no object migration will take place. The transfer of objects between classes improves the division by minimizing the variance within groups. The design of the algorithm, enabling the multiple verification of the validity of the object assignment, reduced the risk of errors. Hierarchical Clustering was characterized by irreversibility: the object assigned to the class remained in it until the end of the analysis. Any error committed at some point could no longer be corrected. In the case the KMCA method, the reorganization at each step allows to correct any assignment. KMCA is a powerful and fast method, which is suitable for large amounts of data. However, it is worth noting that the analysis with *k-means* methods is characterized by a lack of repeatability, which is primarily due to a random location of initial (prototype) classes.

An example of a KMCA analysis of a Raman image of a single cell is shown in Figure 5.7. For calculating the distance (object – centroid), the Manhattan (city block) method was used.

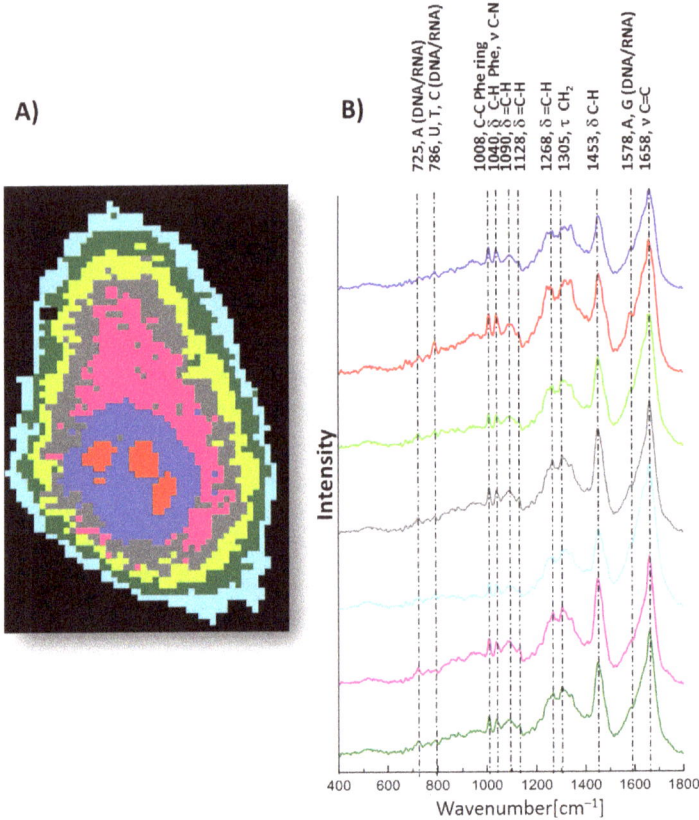

Fig. 5.7 Map of distribution of classes (A) along with their average spectra (B) obtained as a result of the *k-means* cluster analysis (KMCA) performed with the use of Manhattan distance (method for calculating the distance). The measured cell is an endothelial cell from a line HAoEC. Measurement parameters: 488 nm laser line, the integration time 0.65s, size of the imaged area 38×65 μm², the spatial resolution: 0.8 microns (image size: 48×81 points)

Non-hierarchical analysis can also be carried out using Fuzzy *k-means* Cluster Analysis, Fuzzy clustering (FCA). FCA is an iterative method in which objects can be assigned to more than one class. This blurring is based on assigning each object/spectrum to every class at the same time, but with a probability /weight between 0 and 1, where 0 means no membership (lack of any similarity) and 1 – the total membership. [1] An example of a FCA analysis of a Raman image of a single cell is shown in Figure 5.8.

In FCA analysis the membership functions are normalized so that they would sum up to 1. The biochemical composition of every cluster is evaluated by examining the spectrum of its centroid. It is therefore a method similar to *k-means* analysis. Both, FCA images of similarity (single class; Fig. 5.8A) and the final image after the assembly of the similarity maps (Fig. 5.8B) are color-coded.

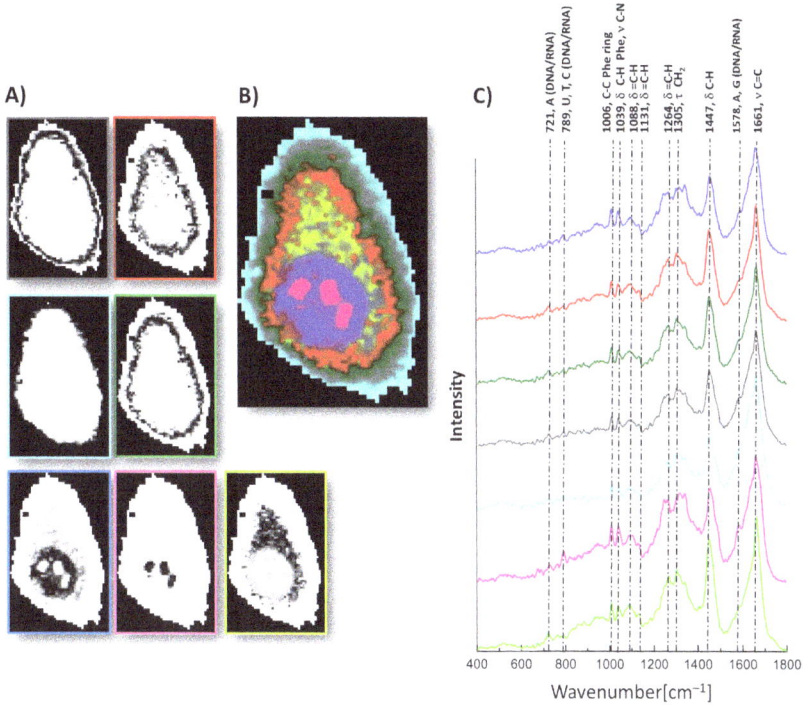

FIG. 5.8. Map of distribution of classes separately (A) and jointly (B) along with their average spectra (C) obtained as a result of the fuzzy cluster analysis (FCA) performed with the D-Values distance (normalised Pearson correlation coefficient) as the method for calculating the distance. The measured cell is an endothelial cell from a line HAoEC. Measurement parameters: 488 nm laser line, the integration time 0.65s, size of the imaged area 38×65 μm^2, the spatial resolution: 0.8 microns (image size: 48×81 points)

5.2.3. Comparison of HCA, KMCA and FCA methods

Comparing the results obtained by HCA, KMCA and the FCA clustering methods, it is clear that the obtained resulting images (Fig. 5.9) differ from each other. This is particularly noticeable and important in the context of the isolation of subcellular structures.

In the examples presented on Figures 5.6 – 5.8 (and summarized in Figure 5.9) each time 7 classes were selected. This allowed to visualize in a satisfying way the major organelles and to obtain averaged spectra that show significant differences. However, some methods did not enable the full identification of all the intracellular structures of the cell. KMCA and FCA (compared to HCA) analysis allowed to distinguish a new class. This class distribution and the profile of the averaged spectra indicate that the new class represents the nucleoli. These structures have not been identified in the hierarchical approach (HCA), even with the increased number of classes, which suggests a lower sensitivity analysis HCA. Additionally,

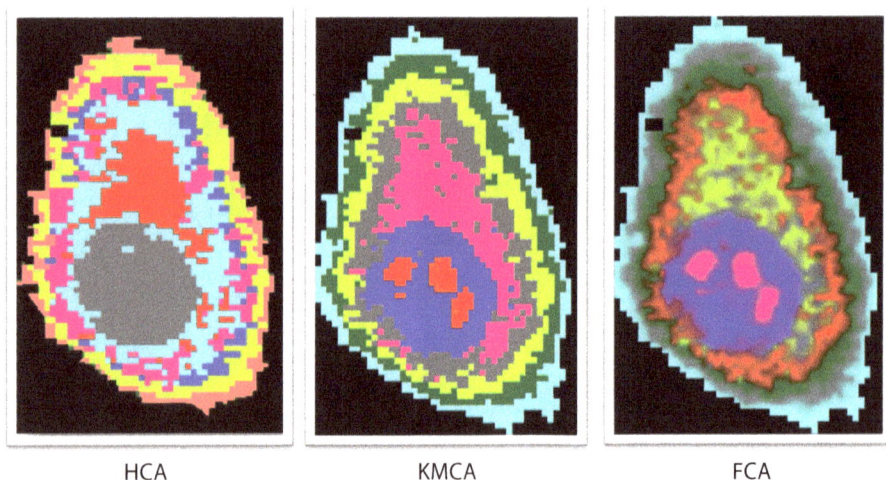

| HCA | KMCA | FCA |

FIG. 5.9. The results of cluster analysis performed with the use of HCA, KMCA and FCA algorithms on spectra obtained from Raman imaging of endothelial cell (experimental details in Fig. 5.6-5.8)

grouping of pixels located in the peri-nuclear area of the cytoplasm is slightly different between the presented methods of cluster analysis. HCA method was proven to indicate better the subtle differences between the spectra, e.g. in the protein and lipid composition (*i.e.* in cytoplasm), while the KMCA and FCA were more sensitive to large differences (due to the fact, that both of these methods are based on centroids).

References

1. Hedegaard M., Matthäus C., Hassing S., Krafft C., Diem M., Popp J., *Spectral unmixing and clustering algorithms for assessment of single cells by Raman microscopic imaging,* Theor. Chem. Acc., **130**, 1249 (2011).

Selected applications of FT-IR spectroscopy

6.1. An effect of molecular symmetry and isotopic substitution on IR and Raman spectra of chloromethane derivatives

Kamilla Malek, Katarzyna M. Marzec

FT-IR absorption and Raman spectroscopy are techniques which enable the determination of molecular structure of chemical species. Both are sensitive to even small changes in structure, resulting in bands shift originating from vibrations of functional groups whose geometric parameters are altered due to chemical reaction, substituent exchange or stress conditions (e.g. temperature, pressure). This application strictly arises from the fact that both spectroscopic methods are complementary. Complementarity of IR and Raman spectroscopy is illustrated in the best way by the principle of mutual exclusion, which says that for a molecule having a center of symmetry, IR-active vibrations are inactive in Raman spectrum and *vice versa*. Firstly one should consider the symmetry of the molecule (so its possible molecular structure) and its impact on the spectral profile on the basis of group theory, since the 3N-6(5) of normal modes of N-atomic molecules can be classified according to its characteristic symmetry point group [1]. For example, a molecule of H_2O with the C_{2v} point group has two vibrations of the A_1 symmetry and one vibration of the B_2 symmetry. Then, the group theory allows predicting a number of active vibrations in the IR and Raman spectra. The selection rules of both spectroscopies indicate that oscillation is active in IR and Raman, if one of the components of dipole moment or polarizability, respectively, has the same symmetry as a specific vibration. It should be also noted that a number of active vibrations may be different than a number of bands observed in spectrum. This can result from the presence of overtones, combination modes, Fermi resonance, lifting of degeneration of some vibrations due to reduction of symmetry or the presence of isomers. Section 1.3 describes normal modes and their symmetry while Section 2.2 describes how to determine the symmetry of vibrations from depolarization ratio in Raman spectra.

Nowadays, it is difficult to imagine the identification of molecular structure without the use of quantum-chemical methods (e.g. using Density Functional

Theory, DFT). This approach dedicated simulation of vibrational spectra is based on the following stages:

- the construction of a model or several structural models of a chosen symmetry for a given molecule,
- optimization of their geometries and identification what structure shows the lowest total energy,
- calculation of frequencies of normal modes, their symmetry, IR intensity and Raman activity, depolarization ratio in the harmonic approximation,
- a comparison of experimental and theoretical spectra for the optimal model,
- a description of the experimental bands based on potential energy distribution (PED) or on visualization of individual vibrations modes for the model selected in the previous step.

An example of such an approach is described in the literature [2,3]. It is important to note that quantum-chemical calculations are limited and their results differ from experimental results because they are conducted for an isolated molecule in vacuum by imperfect calculation methods and basis sets and the lack of anharmonicity.

IR and Raman bands can be assigned to specific vibrations of the functional groups by using the isotope shift. A replacement of one of isotopes by other in a given functional group results in shifting of its IR and Raman bands according to Eq. 1.6 in Chapter 1. This approach is particularly helpful in spectroscopic studies on macromolecules of biological function [4].

AIM OF THE EXPERIMENT

1. The determination of molecular structure from vibrational spectra and group theory.
2. The correlation of experimental results and quantum chemical calculations.
3. An analysis of isotopic shift for a series of chloroderivatives of methane (Fig. 6.1.1).

SCIENTIFIC BACKGROUND

1. Fundamentals of IR and Raman spectroscopy; Chapters 1 and 2.

| carbon tetrachloride | chloroform | dichloromethane |
| symmetry T_d | symmetry C_{3v} | symmetry C_{2v} |

Fig. 6.1.1. Molecular structure of chloroderivatives of methane

EQUIPMENT, MATERIALS, CHEMICALS

1. Liquid CCl_4, $CHCl_3$, CH_2Cl_2 and deuterated derivatives $CDCl_3$, CD_2Cl_2.
2. FT-IR and Raman spectrometers.

PROCEDURE

1. Record Raman spectra of all liquid samples placed in glass ampoules with a spectral resolution of 4 cm^{-1}. Optionally register polarized Raman spectra with the parallel and perpendicular polarization components.
2. Register FTIR spectrum of the samples with the use of the cuvette to measure the liquid samples with a spectral resolution of 4 cm^{-1}.

REPORT

1. Indicate the symmetry elements and the symmetry point groups for CCl_4, $CHCl_3$ and CH_2Cl_2.
2. From group theory calculate irreducible representations for each molecule.
3. Determine the relationship between the symmetry of the molecules and the number of bands active in IR and Raman spectra. Comment the principle of mutual exclusion for CCl_4.
4. Collect bands positions in experimental IR and Raman spectra. For polarized Raman spectra: a) determine intensity in maximum of bands and calculate their depolarization ratio ρ, b) suggest the symmetry of the bands and type of normal modes on the basis of r values.
5. Calculate the experimental isotopic shift for $CDCl_3$ $CHCl_3$ and CH_2Cl_2 and CD_2Cl_2 and assign each band to a vibration of a functional group.
6. Compare experimental data (wavenumber, relative band intensities, symmetry of vibrations, depolarization ratio and isotope shift) with simulated IR and Raman spectra (see Table 6.1.1). Comment differences.

References

1. Cotton, F.A., *Teoria grup: zastosowania w chemii*, PWN, Warszawa, 1973.
2. Marzec, K.M., Gawel, B., Zborowski, K.K., Lasocha, W., Proniewicz, L.M., Malek, K., *Insight into coordination of dilead unit by molecules of 4-thiazolidinone-2-thione. Structural and computational studies*, Inorg. Chim. Acta, **376**, 581 (2011).
3. Marzec, K.M., Reva, I., Fausto, R., Malek, K., Proniewicz, L.M., *Conformational space and photochemistry of* a-*terpinene*, J. Phys. Chem. A, **114**, 5526 (2010).
4. Mak, P.J., Denisov, I.G., Victoria, D., Makris, T.M., Deng, T., Sligar, S.G., Kincaid, J.R., *Resonance Raman detection of the hydroperoxo intermediate in the cytochrome P450 enzymatic cycle*, JACS, **129**, 6382 (2007).
5. Michalska, D., Wysokiński, R., *The prediction of Raman spectra of platinum(II) anticancer drugs by density functional theory*, Chem. Phys. Lett., **386**, 95 (2004).
6. Frish, M.J., et al., *Gaussian 03* Gaussian Inc., Pittsburgh PA, 2003.

Table 6.1.1. Band positions (in cm^{-1}) with symmetry of vibration, relative intensities of the bands in the IR (I_{IR}) and Raman (I_R, converted like in [5]) and depolarization ratio (ρ) for carbon tetrachloride (CCl_4), chloroform ($CHCl_3$), and dichloromethane (CH_2Cl_2) calculated by using B3LYP functional and 6-311++G(d,p) basis set [Gaussian 03 software, 6]

CCl_4 with T_d symmetry

Mode	Band position [cm^{-1}]	Symmetry	I_{IR}	I_R	ρ
1	216	E	0,0	0,46	0,75
2	216	E	0,0	0,46	0,75
3	308	T_2	0,01	0,45	0,75
4	308	T_2	0,01	0,45	0,75
5	308	T_2	0,01	0,45	0,75
6	438	A_1	0,0	1,00	0,0
7	718	T_2	1,00	0,59	0,75
8	718	T_2	1,00	0,59	0,75
9	718	T_2	1,00	0,59	0,75

$CHCl_3$ and $CDCl_3$ with C_{3v} symmetry

Mode	Band position [cm^{-1}]		Symmetry	I_{IR}		I_R		ρ
	$CHCl_3$	$CDCl_3$		$CHCl_3$	$CDCl_3$	$CHCl_3$	$CDCl_3$	
1	256	253	E	<0,01	<0,01	0,88	0,90	0,75
2	256	253	E	<0,01	<0,01	0,87	0,89	0,75
3	358	353	A_1	<0,01	<0,01	1,00	1,00	0,22
4	649	628	A_1	0,03	0,03	0,63	0,68	0,01
5	714	693	E	1,0	1,0	0,16	0,18	0,75
6	714	693	E	1,0	1,0	0,16	0,18	0,75
7	1220	902	E	0,14	0,42	0,11	0,09	0,75
8	1220	902	E	0,14	0,42	0,11	0,09	0,75
9	3055	2249	A_1	<0,01	<0,01	0,17	0,19	0,26

CH_2Cl_2 i CD_2Cl_2 with C_{2v} symmetry

Mode	Band position [cm^{-1}]		Symmetry	I_{IR}		I_R		ρ
	CH_2Cl_2	CD_2Cl_2		CH_2Cl_2	CD_2Cl_2	CH_2Cl_2	CD_2Cl_2	
1	276	272	A_1	<0,01	<0,01	1,00	1,00	0,53
2	685	657	A_1	0,08	0,09	0,66	0,67	0,08
3	705	679	B_2	1,0	1,0	0,18	0,23	0,75
4	884	698	B_1	0,01	<0,01	0,05	0,07	0,75
5	1155	821	A_2	0,0	0,0	0,20	0,18	0,75
6	1276	956	B_2	0,32	0,66	0,09	0,03	0,75
7	1428	1048	A_1	0,0	0,0	0,14	0,14	0,45
8	2997	2172	A_1	0,05	0,05	0,24	0,26	0,09
9	3073	2290	B_1	<0,01	<0,01	0,13	0,14	0,75

6.2. Peak-fitting process by an example of ATR FT-IR spectra of soft tissues

Marlena Gąsior-Głogowska, Adam Oleszko

6.2.1. Skin

The skin is the largest organ of the human body. For the adult, the skin has a surface area of 1.5–2 m² and weighs about 4 kg (without the hypodermis). The thickness of the skin varies from 0.5 mm thick to 5 mm according to anatomical site, age and gender. The skin performs many essential functions: protects internal organs from the outside environmental influences, deleterious radiations, toxic substances, microorganisms and mechanical damage. The skin also performs a secretory and thermoregulatory function. It is a sensory system. The skin helps to maintain the water balance.

The skin has multiple layers, consists of the epidermis (thickness 50–120 μm), the dermis (1–4 mm) and the hypodermis (Fig.6.2.1). An oil-water suspension and an exfoliated keratin cover the surface of the skin, creating so-called the lipid coat of the skin.

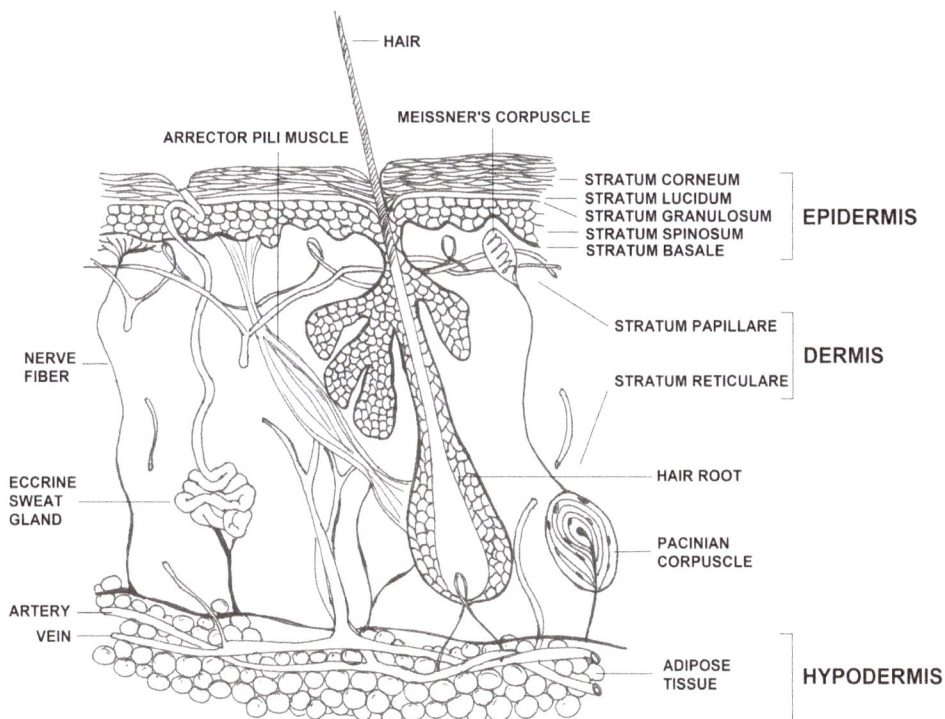

Fig. 6.2.1. Structure of the mammalian skin

Research is often conducted using tissues taken from domestic pigs (lat. *Sus scrofa f. domestica*). Porcine skin is often considered an ideal model for human skin [1, 4].

6.2.2. Skin's proteins

The skin is composed primarily of extracellular matrix (ECM) proteins: collagen and elastin. ECM proteins are present in the dermis, while keratin is found in the epidermis. The hypodermis consists of collagen fibres and adipose cells. Adipocytes store triglycerides [1-3].

Keratins are fibrous proteins produced in epidermal cells called keratinocytes. Their structure is stabilized by numerous disulphide bonds. Keratin is also the key component of accessory structures *i.e.* hair and nails [5]. Collagen is the principal protein in the skin, making up 75 percent of its dry weight, wherein the collagen content changes with age. The most abundant type of collagen in adult skin is type I collagen. Type I collagen composes approximately 80% of the dry weight of skin. Other types of collagen in the skin are collagen type III (15% of total collagen content, 60% of the collagen in foetal skin) and V, VI, VII, XII and XIV. Collagen types I, III and V co-form fibrils with diameter of 25-400 nm. All of the collagen family proteins contain three-stranded helical segments. The right handed triple helix is unique for collagen. Aside from collagen, elastin has been found to make up 1% to 2% of dry weight of the skin [1, 4-8].

6.2.3. Infrared spectroscopy of tissues

In ATR FT-IR measurements the information is collected from the surface of the sample – the technical details are described in Chapter 3.1. The ATR FT-IR spectrum of the skin is dominated by vibrational bands of keratin and epidermal lipids (Table 6.2.1). Weaker bands, originating from other components of the tissue, are also observed: for example bands at 1338 and 1206 cm^{-1} are assigned to collagen. A detailed description of bands in FT-IR spectra is presented in Chapter 6.7.

The maximum peak position of amide I band at 1626 cm^{-1} is typical for proteins with β-sheet conformation. The amide band broadening and the presence of a weaker band at 1651 cm^{-1} indicate the existence of α helical structures.

More detailed information about skin's protein structure can be obtained from the peak-fitting of the amide I band. The peak-fitting (*alias* curve fitting) is the process of decomposition of a complex band into a sum of elementary peaks. Gauss and Lorentz analytical functions (or their sum or linear combination) are used the most often [11]. This method allows estimating the number of overlapping bands and their positions. More detailed information, such as the peak position, full width at half maximum (FWHM), intensity and integrated area of

Table 6.2.1. Major positions and tentative assignments of vibrational modes identified in ATR FT-IR spectra of the skin, adapted from [9-10]

Band position [cm⁻¹]	Assignments
3314 vs br	water; ν(OH); amide A; ν(NH)
2930 w, sh	ν(CH₂)
2852 vw, sh	ν(CH₂)
1736 vw sh	ν(C=O)
1651 sh	amide I; α-helix
1626 s	amide I; β-sheet, β-turns; water δ(OH)
1556 s	amide II
1455 m	δ(CH₂)
1402 m	δ(C(CH₃)₂)
1338 m	δ(CH₂);
1283 m	δ(CH₂); amide III
1242 m	amide III
1206 w	
1081 vw	ν(PO₂⁻), ν(C-O)
1047 w sh	ν(C-O)

Band: vs – very strong, s – strong, m – medium, w – weak, vw – very weak, sh – shoulder, br – broad. Mode: ν – stretching, δ – bending.

each band, can be obtained in this way as well. The peak-fitting process is very often called "deconvolution", but they are completely different mathematical procedures. Deconvolution is widely used to improve spectral resolutions [12-15].

The peak-fitting process of ATR FT-IR spectrum is usually preceded by the data pre-processing *i.e.* atmospheric correction, ATR correction, smoothing, baseline correction and normalization [12–14]. The data pre-processing procedure is performed to reduce the noise and improve SNR (Signal-to-Noise ratio). Most modern spectrometry software tools allow for automated, real-time atmospheric correction during the acquisition of a spectrum. For further processing and analysis of the spectra, commercial spectroscopy software (e.g. OPUS, Origin Pro) is widely used. Another solution is to use some programming environment that allows the users to implement their own algorithms (e.g. Matlab).

Nowadays, smoothing has become a standard procedure and many different algorithms are employed. Common methods use „window" filters (e.g. mean, moving mean, running median, local polynomial regression etc.) and Fourier-transform or continuous wavelet transform. The most popular one is a Savitzky–Golay algorithm. Firstly, a least squares fit of a small set of consecutive data points to a polynomial is performed. Subsequently, the calculated central point of the fitted polynomial curve is taken as the new smoothed data point. A proper selection of window size is crucial. Usually a number of data points which are averaged at once between 5 and 25. The polynomial order is also very important. An incorrect selection of these parameters may lead to disturbances and loosing important information [12].

Background variation is a frequent problem in infrared spectroscopy. It is caused by unequal absorption of the radiation within the sample, light scattering or changing experimental conditions. Hence, a background correction of FT-IR is performed. The simplest method for baseline correction is an offset correction, involving the subtraction of the background approximated with a linear function from the original spectrum. The background can be approximated also with the polyline at the points indicated manually (local minima are commonly used for it). An alternative method is using polynomial baseline correction [13–15].

Prior to a comparative analysis of spectral data series normalization is required. Normalization is commonly done by dividing the original spectrum by intensity (absorbance) or an integral intensity of the specified band – an internal standard. Usually the amide I band is used as the internal standard for normalization, assuming no quantitative changes of proteins. For a simple band, Min-Max normalization is widely applied. In this method, the original spectrum is scaled between zero and one that is the maximal absorbance value of the spectrum in the selected

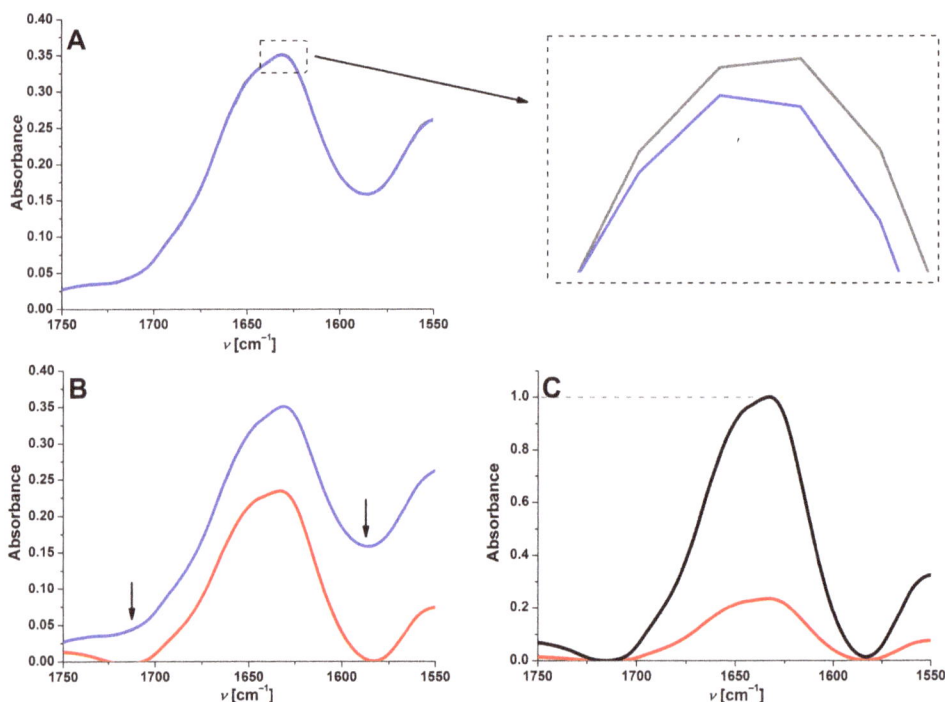

Fig. 6.2.2. Consecutive spectral processing steps of ATR FT-IR spectrum of the skin in the amide I range: A – spectral smoothing; B – background subtraction; C – normalization. The grey line indicates the original unprocessed spectrum. Spectrum were smoothed with Savitzky-Golay filter with the window width of 15 (blue line) and the baseline were corrected (red line). The arrows indicate places of the baseline subtraction. Finally, the spectrum was normalized using Min-Max method (black line)

spectral region equals one, the minimum zero. Another way to normalize the spectrum is through vector normalization. First, the average value of the absorbances is calculated for the normalized spectral region. This value is subtracted then from the spectrum such that a new average value equals zero and the sum squared deviation equals one [14–15].

Mid-infrared spectra of tissues are very complex and they consist of the sum of all signals originating from vibrations of their biomolecules. In consequence, FT-IR spectra of biological samples may be difficult to interpret. Hence, decomposition of a complex band into sub-bands is often carried out. For this purpose, the number of overlapping bands and their wavenumber positions are determined by calculating the second derivative of the spectrum (Fig. 6.2.3). The minima of the second derivative spectrum give the positions of the component bands.

An exemplary curve-fitting of the amide I band in the pig skin spectrum is illustrated in Figure 6.2.4. The peak-fitting of ATR FT-IR skin spectrum in the region of 1700-1600 cm^{-1} allowed for separation of ten components arising from the different types of secondary structures present in tissue proteins. The most intensive feature at 1649 cm^{-1} is assigned to α-keratin. The component band at 1637 cm^{-1} originates from β-keratin and bending vibrations of the O-H group of tissue. Intensities of bands at 1692, 1670, 1660 and 1628 cm^{-1} are affected by amide bond vibrations of collagen. The presence of these bands and their positions are typical of that protein family [16].

AIM OF THE EXPERIMENT

1. An analysis of ATR FT-IR spectra of complex materials, *i.e.* tissue.
2. Learning peak-fitting methods.

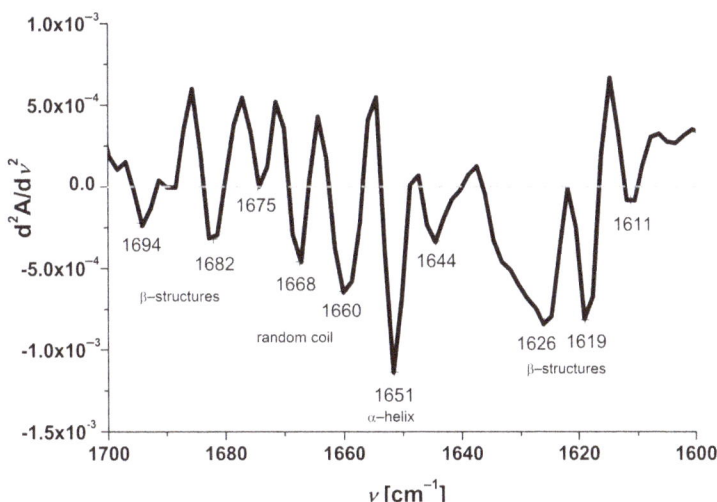

Fig. 6.2.3. Second derivative of ATR-FTIR spectrum of domestic pig' skin in the amide I region (1700-1600 cm^{-1})

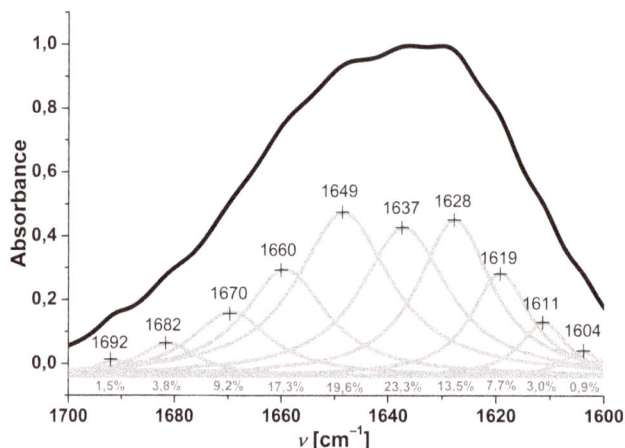

Fig. 6.2.4. Peak-fitting of pig skin ATR FT-IR spectrum in the amide I region. Bands positions and percentage area obtained after curve-fitting are given

SCIENTIFIC BACKGROUND

1. Fundamentals of FT-IR spectroscopy; Chapter 1.
2. Construction and operation of FT-IR spectrometer, including the ATR technique; Chapter 3.1.
3. Basic knowledge of secondary structures of proteins and their marker bands in FT-IR spectra; Chapter 6.7.

EQUIPMENT, MATERIALS, CHEMICALS

1. Soft tissue specimens, i.e. skin or tendons obtained from farm animals (swine, rabbits, rats). Lyophilised bovine collagen as an exemplary reference material.
2. FT-IR spectrometer equipped in ATR, e.g. Bruker ALPHA.
3. OPUS, ORIGIN Pro 9.0, Matlab or other software with curve fitting algorithms built in.

PROCEDURE

1. Collect of ATR FT-IR spectra of soft tissue and collagen, including optimizing measurement parameters (spectral resolution – 4 cm^{-1}, 128 scans for background and 32 for sample, spectral range 4000-400 cm^{-1}).
2. Identify marker bands of tissue structural components, according to Table 6.2.1. Indicate amide bands.
3. Perform the curve-fitting of obtained spectra in amide I range, preceded by spectral data pre-processing and calculation of second derivative spectra – according to methods described in Chapter 6.2.2.
4. Determine positions of overlapping bands in the amide I range and assign them to the type of secondary structure of proteins, see Chapter 6.1.1.
5. For each component, calculate the percentage of sub-band area (integrated intensity) over the total amide I band area.

6. Indicate the dominant component of the amide I band in spectra of studied tissues and the reference protein.

   REPORT

1. Indicate main structural proteins by peak-fitting of ATR FT-IR spectra of analysed soft tissues in the amide I spectral region.
2. Discus the utility of the curve-fitting method to study IR spectra of tissues.

References

1. Gąsior-Głogowska M., Zastosowanie spektroskopii Ramana do monitorowania zmian strukturalnych białek w tkankach miękkich poddanych rozciąganiu. Doctoral Dissertation, Wroclaw University of Technology, Wrocław 2013.
2. Mescher A.L. (ed.), *Junqueira'S Basic Histology: Text and Atlas*, McGraw-Hill, USA 2013.
3. Busam K.J. (ed.), *Dermatopathology*, Saunders Elsevier, USA 2009
4. Meyer W., Neurand K., Radke B., *Collagen fibre arrangement in the skin of the pig*, J. Anat,. **134**, 139 (1982).
5. Bragulla H.H., Homberger D.H., *Structure and functions of keratin proteins in simple, stratified, keratinized and cornified epithelia*, J. Anat., **214**, 516 (2009).
6. Kłyszejko-Stefanowicz L., *Cytobiochemia. Biochemia niektórych struktur komórkowych*, WN PWN, Warszawa 1995.
7. Banaś M., Pietrucha K., *Typy i struktura białka kolagenowego*, Zeszyty Naukowe Politechniki Łódzkiej, Chemia Spożywcza i Biotechnologia, 1058, 93 (2009).
8. Ricard-Blum S., Ruggiero F., *The collagen superfamily: from the extracellular matrix to cell membrane*, Pathol. Biol., **53**, 430 (2005).
9. Greve T.M., Andersen K.B., Nielsen O.F, *ATR-FTIR, FT-NIR and near-FT-Raman spectroscopic studies of molecular composition in human skin in vivo and pig ear skin in vitro*, Spectroscopy, **22**, 437 (2008).
10. Ali S.M., Bonnier F., Lambkin H., Flynn K., McDonagh V., *A Comparison of Raman, FTiR and ATR-FTIR Micro Spectroscopy for Imaging Human Skin Tissue Sections*, Anal. Methods, **5**, 2281 (2013).
11. Iskander R., w: *Chemia fizyczna t.4. Laboratorium fizykochemiczne* (red. Komorowski L., Olszewski A.), WN PWN, Warszawa 2013, p. 81-84.
12. Rinnan A, van der Berg F., Engelsen S.B., *Review of the most common pre-processing techniques for near-infrared spectra*, .Trends Anal. Chem., **28**, 1201 (2009).
13. Lash P., *Spectral pre-processing for biomedical vibrational spectroscopy and microspectroscopic imaging*, Chemometr. Intel. Lab., **117**, 100 (2012).
14. Stuart B., *Infrared Spectroscopy: Fundamentals and Application*, John Wiley & Sons, UK 2004.
15. Smith B.C., *Fundamentals of Fourier Transform Infrared Spectroscopy*, CRC Press, USA 2011.
16. de Campos Vidal B., Mello M.L.S., *Collagen type I amide band infrared spectroscopy*, Micron, **42**, 283 (2011).

6.3. Synthesis and spectral characteristics of hydroxyapatites

Marlena Gąsior-Głogowska, Adam Oleszko

6.3.1. Hydroxyapatites

Hydroxyapatite (HAp) is a name given to calcium orthophosphate with the chemical formula of $Ca_{10}(PO_4)_6(OH)_2$. It is a component of phosphate minerals and biological apatite found in vertebrate bones and teeth. Hydroxyapatite is also the most common form of pathologic calcifications in the body, *i.e.* urinary stones and gallstones. The stoichiometric hydroxyapatite has a Ca/P molar ratio equal to 1.667 and crystallizes in a monoclinic crystal system. The hexagonal structure of natural hydroxyapatite (Fig. 6.3.1) is due to a lack of stoichiometry.

Biological hydroxyapatites are non-stoichiometric. Various isomorphous substitutions, *via* cation or anion exchange, modify the structure of bio-apatites. The PO_4^{3-} ions can be replaced by CO_3^{2-}, SO_4^{2-}, SiO_4^{4-} or AsO_4^{3-} ions. Carbonate ions can substitute hydroxyl or phosphate ions forming carbonate hydroxyapatite of type A or B. Several cations can substitute calcium ions in the apatite structure, e.g. Mg, Na, K, Sr, Zn and other trace elements. The substitutions affect stoichiometry, crystallinity and thermal, chemical and biological stability of apatite. For example, the incorporation of Mg ions into the hydroxyapatite lattice alters the size of crystals and increases the solubility of apatite [2-6].

Hydroxyapatite materials, due to the high level of biocompatibility with bone tissue and biomolecule loading capacity, are widely used in orthopaedic and dental fields, laryngology and cosmetology. Hydroxyapatite for medical applications can be extracted from natural resources (e.g. minerals, bones, corals) but in the most cases, they are produced through synthesis. There are several methods of preparing HAp crystals, including wet chemical deposition (in aqueous solutions or suspensions, with pH-control), dry method (solid state reaction), flux, hydrothermal and sol-gel synthesis. The most commonly used methods of HAp synthesis are wet

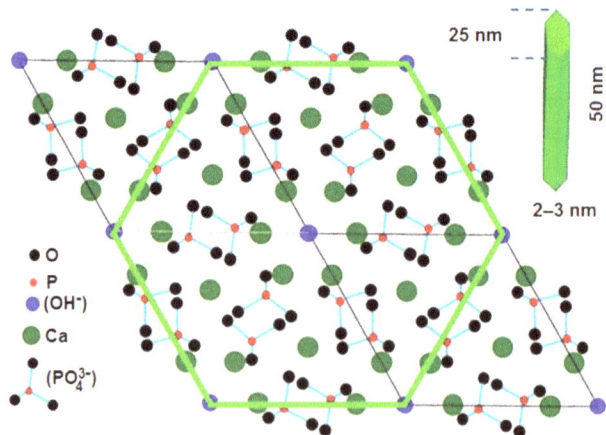

Fig. 6.3.1. Crystalline structure of hydroxyapatite [1]

techniques incorporating foreign ions into the apatite crystal lattice. Apatite materials prepared this way are poorly crystalline or even amorphous [4]. Examples of wet synthesis reactions are listed in Table 6.3.1.

Table 6.3.1. Common reagents used for the synthesis of hydroxyapatite by the wet chemical technique [4]

substrate 1	substrate 2	reaction	pH
$Ca(NO_3)_2$	$(NH_4)_2HPO_4$	$10\ Ca(NO_3)_2 + 6\ (NH_4)_2HPO_4 \rightarrow Ca_{10}(PO_4)_6(OH)_2 + 12\ NH_4NO_3 + 8\ HNO_3$	8-12
$Ca(OH)_2$	H_3PO_4	$10\ Ca(OH)_2 + 6\ H_3PO_4 \rightarrow Ca_{10}(PO_4)_6(OH)_2 + 18\ H_2O$	8
$CaCl_2$	Na_2HPO_4	$10\ CaCl_2 + 6\ Na_2HPO_4 \rightarrow Ca_{10}(PO_4)_6(OH)_2 + 12\ NaCl + 8\ HCl$	
$Ca(NO_3)_2$	$(NH_4)_3PO_4$	$10\ Ca(NO_3)_2 + 6\ (NH_4)_3PO_4 + 2\ NH_3 \cdot H_2O \rightarrow Ca_{10}(PO_4)_6(OH)_2 + 20\ NH_4NO_3$	≥ 9

The shape, size and physicochemical properties of synthetic hydroxyapatite depend on concentrations of reagents, pH of solution and temperature of the reaction.

6.3.2. Infrared spectroscopy of hydroxyapatite

In the case of hydroxyapatite obtained from wet chemical synthesis, the degree of sample drying plays a key role. Because of the strong absorbance of water in the mid-infrared spectral region, infrared spectroscopic studies are performed on dried HAp samples. However, the drying time and temperature influence the final product structure.

FTIR spectra of apatite show characteristic bands due to vibrations of various structural groups (Table 6.3.2). The band at 962 cm^{-1} assigned to the phosphate stretching vibration ($v_1(PO_4^{3-})$) and the strong complex band at ca. 1029 cm^{-1} ($v_3(PO_4^{3-})$) are markers for apatite. The positions and shapes of these bands are affected by the crystallite size and isomorphic substitutions. If monohydrogenphosphate ions (HPO_4^{2-}) are presented in the apatite crystal lattice, an additional sub-band at 1110 cm^{-1} is noted. The two-component bands at about 1030 and 1015 cm^{-1} indicate stoichiometric and non-stoichiometric hydroxyapatite, respectively (Fig. 6.3.2A). Substituted hydroxyapatites have additional bands in the region of 910-850 cm^{-1}. In carbonated apatite, the CO_3^{2-} characteristic band assigned to the carbonate stretching vibration ($v_2(CO_3^{2-})$) appear in their IR spectra. The position of this band differs slightly depending on the type of the carbonate substitution in the crystal lattice. The type B of carbonated apatite is characterized by a v_2 band at ca. 880 cm^{-1}, whereas the type A configuration has this band at about 872 cm^{-1}. The sub-band at 866 cm^{-1} is assigned to labile CO_3^{2-} ions loosely attached to the surface of hydroxyapatite (Fig. 6.3.2B). The carbonate content can be estimated from the ratio of the integrated areas of the bands $v_2(CO_3^{2-})$ and $v_3(PO_4^{3-})$ ($\sim$1029 cm^{-1}) or $v_1(PO_4^{3-})$ ($\sim$960 cm^{-1}) [5-7].

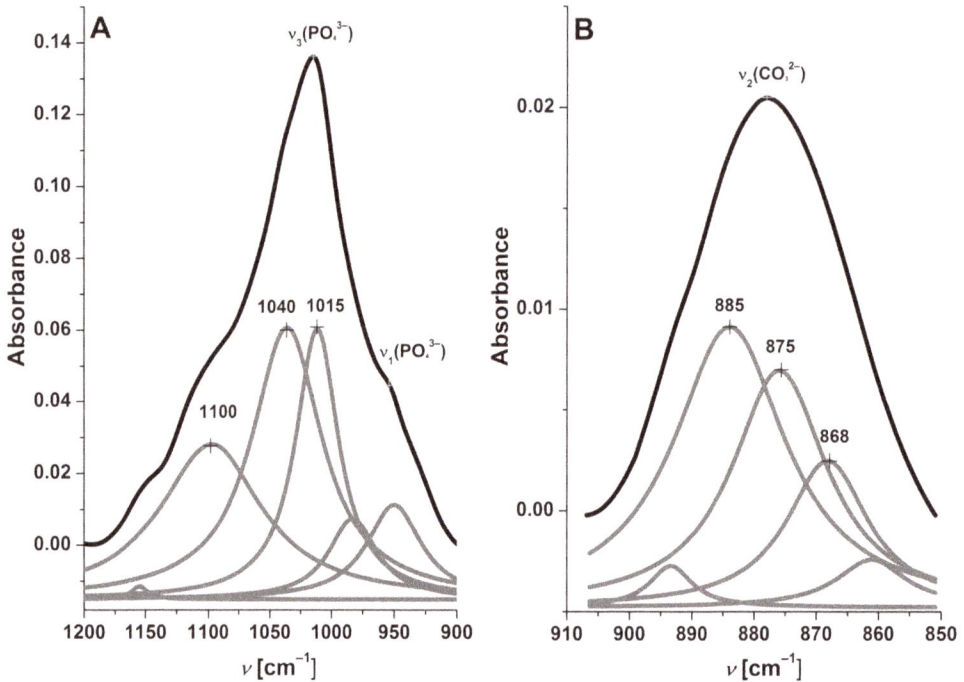

Fig. 6.3.2. The representative curve-fitted ATR FT-IR spectrum of the carbonated-substituted apatite in the range of: A – stretching vibrations (v_1, v_3) of the phosphate groups (1200-900 cm^{-1}); B – bending vibrations (v_2) of the carbonate groups (910-850 cm^{-1})

AIM OF THE EXPERIMENT

1. A qualitative analysis of synthesized hydroxyapatite using infrared absorption spectroscopy.
2. Identification of the HAp type present in biological specimens.

SCIENTIFIC BACKGROUND

1. Fundamentals of FT-IR spectroscopy; Chapter 1.
2. Construction and principle of operation FT-IR spectrometer including ATR technique; Chapter 3.1.
3. Spectral processing; Chapter 6.2.

EQUIPMENT, MATERIALS, CHEMICALS

1. Powdered pig bone sample as a reference material.
2. Reagents for HAp preparation:
   A solution: 11 g $CaCl_2$ (anhydrous) dissolved in 100 mL of distilled water
   B solution: 4.5 g $Na_3PO_4 \cdot H_2O$ dissolved in 100 mL of distilled water
   C solution: 4.5 g $Na_3PO_4 \cdot H_2O$ and 5.5 g Na_2CO_3 dissolved in 100 mL of distilled water distilled water for washing.
3. Heated ultrasonic bath, centrifuge and/or set for filtration under lower pressure, laboratory dryer.

Table 6.3.2. Major positions and tentative assignments of vibrational modes identified in ATR FT-IR spectra of hydroxyapatites, adapted from [5]

Band position [cm^{-1}]		Assignment
HAp	CHAp	
3567 m	>3553 br	structural hydroxyl ions; ν(OH)
3400 br		water; ν(OH)
	1528 w (A)	
	1498 sh (A)	
	1470 s	
	1450 s (B)	
	1410 s (B)	carbonate; ν_3(CO_3^{2-})
1092 vs	1094 s	
1060 sh	1060 sh	phosphate; ν_3(PO_4^{3-}) and carbonate ν_1(CO_3^{2-})
1029 vs	1029 vs	
961,7 s	961,3 m	phosphate; ν_1(PO_4^{3-})
	933 sh	
	880 m (A)	
	873 s (B)	
856 wbr	847 sh	carbonate; ν_2(CO_3^{2-})
	813 vw (B)	
	760 vw (B)	
	668 m (A)	carbonate; ν_4(CO_3^{2-})

Band: vs – very strong, s – strong, m – medium, w – weak, vw – very weak, sh – shoulder, br – broad. Mode: ν – stretch of **hydroxyl groups**; ν_1 – symmetric stretch, ν_3 – triply degenerate asymmetric stretch of **phosphate groups**; ν_1 – symmetric stretch, ν_2 – out of plane bending, ν_3 – asymmetric stretch, ν_4 – doubly degenerate bending of **carbonate groups**. Type of carbonate substitution: (A) – A type, (B) – B type.

4. FT-IR spectrometer equipped in ATR, e.g. Bruker ALPHA.
5. OPUS, ORIGIN Pro 9.0 software.

PROCEDURE

1. Wet synthesis of hydroxyapatite:
   Prepare the solutions A, B and C. Add 50 mL of solution A into solutions B and C. Use a ultrasonic bath (time setting 40 minutes) or a centrifuge (15 mins, 3500 rpm). Concentrate the solution under the reduced pressure, as needed. Double wash obtained precipitates with distilled water. Desiccate precipitates in a lab dryer (24h, 72°C) or allow it to air dry in room temperature for a week.
2. Measure ATR FT-IR spectra of synthesised hydroxyapatites and the bone specimen by following measurement setup: spectral resolution – 4 cm^{-1}, 256 scans for background and 128 for sample, spectral range 4000-400 cm^{-1}.

REPORT

1. Indicate marker bands for hydroxyapatite in ATR FT-IR spectra.
2. Make the peak-fitting process for obtained spectra in the spectral range of 1200–900 cm^{-1}, see Chapter 6.2.
3. Indicate bands corresponding to the carbonate ions incorporated into crystal lattice of obtained HAp and bone, see Chapter 6.3.2.
4. Peak-fit the obtained spectra in the spectral range of 910-850 cm^{-1}, Chapter 6.2. Specify the type of substitution, see Chapter 6.3.2.
5. Estimate the level of carbonate substitution in synthesised apatite and the bone tissue specimen. Use the ratio of the carbonate band integrated area $v_2(CO_3^{2-})$ to phosphate band $v_1(PO_4^{3-})$.

References

1. http://www.iupui.edu/~bbml/boneintro.shtml
2. Wopenka B., Pasteris J.D., *A mineralogical perspective on the apatite in bone*, Mater. Sci. Eng. C, 25, 131 (2005).
3. Sobczak-Kupiec A., Wzorek Z., *Właściwości fizykochemiczne ortofosforanów wapnia istotnych dla medycyny – TCP i HAp*, Czasopismo Techniczne. Chemia, 10, 309 (2010).
4. Sobczak A., Kowalski Z., *Metody mokre otrzymywania hydroksyapatytu*, Czasopismo Techniczne. Chemia, 13, 125, (2008).
5. Antonakos A., Liarokapis E., Leventouri T., *Micro-Raman and FTIR studies of synthetic and natural apatites*, Biomaterials, **28**, 3043 (2007).
6. Pleshko N., Boskey A., Mendelsohn R., *Novel infrared spectroscopic method for the determination of crystallinity of hydroxyapatite minerals*, Biophys. J., **60**, 786 (1991).
7. Boskey A.L., Gadaleta S., Gundberg C., Doty B., Ducy P., Karsenty G., *Fourier transform infrared microspectroscopic analysis of bones of osteocalcin-deficient mices provides insight into the function of osteocalcin*, Bone, **23**, 187 (1998).

6.4. The application of infrared spectroscopy for the determination of petroleum hydrocarbons in surface water and wastewater

Paweł Miśkowiec

Acronyms:

- EPA – The Environmental Protection Agency – US federal agency acting to protect human health and the environment.
- ASTM – American Society for Testing and Materials – the main organization in the United States, which develops standards, used outside the United States as well.
- ISO – International Organization for Standardization – a non-governmental organization, based in Geneva dealing among others with the development of standards.

■ PN – Polish Standard (pol. *Polska Norma*) – a Polish national standard adopted and approved by the Polish Committee for Standardization (PKN).

6.4.1. The composition of crude oil; environment pollution with petroleum products

Petroleum is the main source of alkanes, cycloalkanes, and aromatic hydrocarbons. Its composition is extremely complex and depends on the place of excavation. The largest petroleum producing countries are currently Saudi Arabia, Russia and the USA. Crude oil is separated by distillation into several fractions. Typical and the most important fractions are given in Table 6.4.1.

Crude oil and products of its distillation are a frequent cause of environmental pollution with organic compounds. Leaks from tankers are always ecological disaster (the Exxon Valdez in 1989, the Prestige in 2002). Serious oil spills also occur in the vicinity of mining areas and leaky pipelines both on the land and the sea.

In the case of crude oil processing the most vulnerable to contamination are the surroundings of refineries and petrochemical plants, as well as the areas in the vicinity of paint shops or petrol stations. High concentrations of petroleum products are recorded near car repair shops, car parks and major thoroughfares. Soil and ground water are particularly vulnerable components of the environment. Water in rivers and lakes is also exposed to contamination through the surface runoff.

One of the most important petroleum toxicant from the sources listed above is mineral oil. It is defined as a petroleum distillate, mixture of liquid and solid hydrocarbons having a carbon chain length of from C_{12} to C_{35} and boiling temperature above 200°C. The oil composition varies, depending on the origin and processing technology. Talking about application, mineral oil is widely used in the industry and automotive. The main oil industrial applications are as follow: lubricants, protectors against metal corrosion, cooling and hardening agents, components of cosmetics and medicines, softeners in the production of plastics. Oil is also commonly used as liquid fuel in diesel engines. The solubility in water of the components of mineral oil decreases with increasing molecular mass of alkanes and cycloalkanes

Table 6.4.1. Typical petroleum fractions

Fraction	Distillation temp. (°C)	The number of carbon atoms
Gas	less than 20	C_1-C_4
Petroleum ether	20 – 60	C_5 – C_7
Ligroin (naphtha)	60 – 100	C_6 – C_7
Petrol	40 – 205	C_5 – C_{11} and cycloalkanes
Oil	175 – 250	C_{12} – C_{15}
Diesel	180 – 350	C_{12} – C_{25}
Lubricating oil	non-volatile liquid	cyclic with long branched
Asphalt or petroleum coke	Non-volatile solid	polycyclic structures

in the mixture, but it increases with increasing concentrations of other organic contaminants such as solvents or surfactants. Any uncontrolled leakage of the mineral oil is a serious environmental problem because of its durability and hence the remediation difficulties.

6.4.2. The toxic properties of the components of crude oil

When talking about toxicity of crude oil one have to consider its particular fractions. The lightest fraction is composed of gaseous hydrocarbons, *i.e.* from methane to butane. High concentrations of these gases in the air can cause hypoxia and even death (natural gas poisoning). However, even the exposure to small concentrations of the hydrocarbon gas can cause a narcotic effect. Hydrocarbons with 5 – 8 carbon atoms in the chain (*i.e.* typical gasoline components) are neurotoxic. Neurotoxic properties have been thoroughly tested for n-hexane, which is a typical component of gasoline and commonly used as a solvent. Hexane is oxidized in the organism *via* several enzymatic reactions which can be depicted in the following scheme:

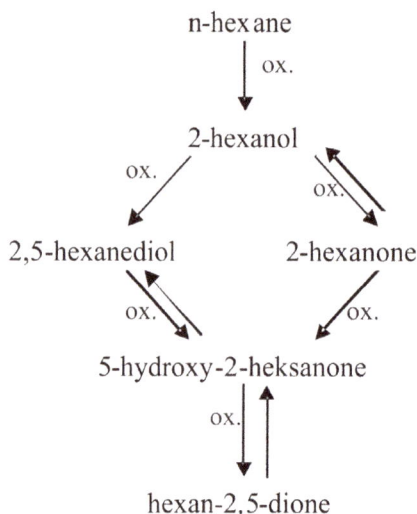

Fig. 6.4.1. Oxidation of n-hexane in the organism

Hexane-2,5-dione (CH_3-CO-CH_2CH_2-CO-CH_3) is one of the strongest known neurotoxicants. The continuous and repeated exposure to oil distillation products (including heavier fractions) may cause damage in the gastrointestinal tract and lungs as well as skin dryness or cracking. Petroleum products are highly lipophilic. Thus, they can dissolve fluid lipids, which lead to the destruction of cell membranes and nerves shells [1,2].

6.4.3. Legislation concerning problems of environmental pollution with petroleum products and methods of its measurement

The limit values of petroleum hydrocarbons in surface water, groundwater and wastewater in Poland is regulated by a number of regulations *(the limit concentration of petroleum hydrocarbons governed by the particular regulation is given in italics):*

1. The regulation of the Minister of the Environment from 22nd of October 2014 – "The classification status of surface water and the environmental quality standards for priority substances." *(class 1st and 2nd of water: ≤ 0.2 mg / L).*
2. The regulation of the Minister of Environment from 27th of November 2002 – "The demands for surface water which is used to supply the population for drinking water *(the water categories: A1: ≤ 0.05 mg/L, A2: ≤ 0.2 mg/L, A3: ≤ 1 mg /L).*
3. The regulation of the Minister of Environment from 18th of November 2014. "On the conditions that must be met while discharging of sewage into water or soil and on the substances particularly harmful to the aquatic environment *(industrial wastewater from oil refineries ≤ 5 mg/L, the other types of industrial wastewater ≤ 15 mg/L).*
4. The regulation of the Minister of Environment from 23rd of July 2008. "On the criteria and method of evaluation of groundwater *(Class $1^{st} \leq 0.01$ mg/L, $2^{nd} \leq 0.1$ mg/L, $3^{rd} \leq 0.3$ mg/L, $4^{th} \leq 5$ mg/L, $5^{th} > 5$ mg/L).*
5. The regulation of the Minister of Construction from 14th of July 2006. "On the methods of realization of duties of industrial waste suppliers and on the conditions of wastewater disposal into the sewage system. *(industrial wastewater disposed to the sewage system: ≤ 15 mg/L).*

The methodology of the analysis of petroleum hydrocarbons in the environment is generally based (both in Poland and in the world) on two techniques: infrared spectroscopy and gas chromatography. Below one can find the most important protocols describing the methodology of quantification of petroleum hydrocarbons in the environment, used in the laboratories:

- EPA 418.1 (petroleum hydrocarbons, IR spectroscopy, 1994)
- ASTM D7066-04 (oils and greases, IR spectroscopy, 2011).
- ASTM D7678-11 (petroleum hydrocarbons in water/wastewater, IR spectroscopy, 2011).
- ISO/TR 11046 (mineral oil in the soil, IR spectroscopy, gas chromatography, 1994);
- ISO 9377-2 (mineral oil in water, gas chromatography, 2000)
- PN-82/C-04565.01 (Non-polar aliphatic hydrocarbons from crude oil in water and wastewater, IR spectroscopy, 1982).

Since 2006 two protocols are used interchangeably in the Polish laboratories: based on infrared spectroscopy (PN-82/C-04565.01) and gas chromatography

(PN-EN ISO 9377-2). The main reason for the development of chromatographic techniques as the reference method was the introduction of strict record of consumption of the main extractant in the spectroscopic technique – carbon tetrachloride, which is known as highly toxic and destructive for the ozone layer.

However, spectroscopy methods have several advantages and from this reason they are still widely used in laboratories. First of all this is a rapid and relatively inexpensive method of the determination of total petroleum hydrocarbon content in water. The spectroscopic analyses can be performed using a relatively small size and low-cost spectrometers. To cope with the problem of extractant tetrachloroethylene C_2Cl_4 can be used instead of harmful CCl_4, as a much less toxic and harmless to the ozone layer.

Usually transmission FT-IR technique is used for the above-mentioned identification and quantification of oil. However, reflection technique ATR can be applied as well in the analysis of oil in water. In this case, the extraction can be done using a volatile organic solvent (e.g. n-hexane). After solvent evaporation the residue is a subject of spectroscopic analysis. The disadvantage of this method is the possibility of removal a part of the analysed hydrocarbons with the evaporating solvent. When a sample prepared in the form of a layer has a thickness less than the depth of penetration of the beam of infrared radiation, the entire analyte participates in absorption of radiation and absorbance *versus* concentration values follows the Lambert-Beer law. This method is described in detail in [3].

6.4.4. An application of IR spectroscopy in analytical chemistry

Infrared spectroscopy is mainly used to identify compounds and to examine their structures. However, knowing that the absorption of IR radiation typically occurs in accordance with the law of Lambert – Beer, infrared spectroscopy can be used as a quantitative analytical method.

Hydrocarbon molecules have the following moieties of carbon atoms linked to hydrogen atom: $-CH_3$ (the methyl group and the primary carbon atom); $-CH_2-$ (the methylene group, the secondary carbon atom), $-CH<$ (a tertiary carbon atom, in a place of chain branching). Stretching vibrations of carbon-hydrogen bonds occur in the range of $2700 – 3300$ cm^{-1}. Depending on the carbon order the absorption bands appears at a slightly modified wavenumber. Figure 6.4.2 presents a spectrum of a typical alkane in this region.

The most characteristic band of the $-CH_2-$ groups (asymmetrical stretching, v_{as}) occurs at the wavenumber around $2926–2930$ cm^{-1}. The absorbance of this band is proportional to the number of $-CH_2-$ groups and therefore to the concentration of hydrocarbons containing this moiety. The analysis of this band is the basis of the quantitative determination of petroleum hydrocarbons.

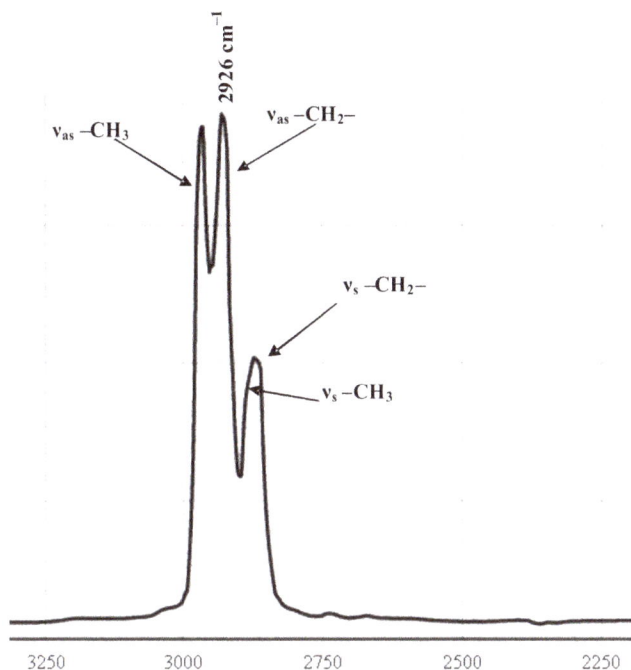

Fig. 6.4.2. A typical IR spectrum of alkane in the range 2250–3250 cm^{-1}

6.4.5. Nernst distribution law

As in this exercise, an extraction of petroleum hydrocarbons from the aqueous phase to strongly hydrophobic – tetrachlorethylene is necessary; some basic facts about the extraction process between two immiscible substances are discussed below. For more details the reader is referred to [4].

Let's consider a system of two separate volumes of immiscible liquids 1 and 2 (the absolute immiscibility is assumed), e.g. water and tetrachlorethylene. In this system we dissolve a certain substance Z. This substance is soluble in both liquids 1 and 2. Its concentration in the liquid 1 is x_1 and in the liquid 2 x_2. According to the Nernst law, concentrations x_1 and x_2 are related to the following relationship: $\frac{x_2}{x_1} = K(T)$. The constant K is called the partition coefficient or the distribution coefficient. The notation $K(T)$ reminds that the K factor is a function of temperature and when giving its value, temperature of determination should be added too. The Nernst's distribution law is the basis of the extraction process, which is often used in the chemical industry and analytics for separating mixtures and purification of reaction products. In the case of this exercise, the liquid 2 is tetrachlorethylene while the liquid 1 is water and the determined hydrocarbons are the solute Z. More than 99% of the measured hydrocarbons pass during the extraction of the aqueous phase to tetrachloroetylene phase.

AIM OF THE EXPERIMENT

The aim of the exercise is to gain skills of liquid sample preparation and collection of their FT-IR spectra (including anhydrous solutions), and the determination of petroleum hydrocarbons in water.

The exercise is based on the method described in the Polish Standard PN-82 C-04565.01 *Determination of the content of crude oil and its components. Determination of non-polar aliphatic hydrocarbons with infrared spectroscopy* [5]. The method described below is used to determine the sum of non-polar aliphatic hydrocarbons in water and sewage contaminated with mineral oil. This quantification relies on the extraction of organic compounds from a sample with tetrachlorethylene or different halogen derivatives, then the separation of polar compounds by their adsorption on a silicon oxide or aluminium oxide, and finally the quantitative determination of the remaining non-polar aliphatic hydrocarbons with infrared spectroscopy in the wavenumber range of 3200–2700 cm^{-1}. The measure of the content of the mineral oil is the value of absorbance at a wavenumber 2926 cm^{-1}, depending on the number of -CH_2- groups.

It should be noted that by using this method also the fraction of lighter hydrocarbons (C7-C11) can be extracted. However, they are volatile enough to evaporate before sampling unless the sampling conditions results of the sudden leakage of gasoline from tanks.

SCIENTIFIC BACKGROUND

1. Fundamentals of IR absorption spectroscopy; Chapter 1,
2. Construction and principles of operation of a FT-IR spectrometer, FT-IR techniques; Chapter 3.1.

EQUIPMENT, MATERIALS, CHEMICALS

1. Tetrachlorethylene or other organic solvent without -CH_2 and -CH_3 functional groups, spectrally pure.
2. Mineral oil for IR spectroscopy, or any alternative mixture of alkanes.
3. The silicon oxide or aluminium oxide neutral, for chromatography.
4. Samples of water/wastewater contaminated with mineral oil.
5. FT-IR spectrometer.
6. Software for spectra processing, *i.e.* OPUS, OMNIC, Origin.

PROCEDURE

I. Preparation of standard solutions and calibration curve.
   One can use mineral oil for IR spectroscopy as a source of saturated hydrocarbons C15 to C50. An alternative mixture of alkanes is acceptable.
   a. Standard solution I.
      Add 100 mg of the above mentioned standard to a 25 mL volumetric flask. Make up to the mark with tetrachlorethylene and mix thoroughly. Calculate the content of the standard in 1 mL of the solution.

b. Standard solution II.

Add such a volume of standard working solution I to a volumetric flask of 25 mL to obtain solution containing 0.4 mg of standard in 1 mL of solution. Make up to the mark with tetrachlorethylene and mix thoroughly.

c. Preparation of the calibration curve.

Pipette sequentially 0.0 µL; 50 µL; 100 µL; 200 µL, 500 µL, 1000 µL of working solution II into the six 10 mL flasks. Make up every flask to the mark with tetrachlorethylene and mix thoroughly. Calculate concentration of each standard in mg/mL. Transfer every standard into a cuvette suitable for IR spectra measurements. Record the spectrum in the range of 2500-3500 cm^{-1} with a spectral resolution of 2 cm^{-1}. Baseline of each spectrum should be corrected. Record absorbance values at wavelength of 2926 cm^{-1}. Collect the results in the table below:

Concentration of standard solution [mg/mL]	A (2926 cm^{-1})

Plot the absorbance versus the concentration of standards, fit a line using linear regression method, and calculate the equation and determination coefficient R^2.

II. Determination of petroleum hydrocarbons in water

Measure up to 500 mL of each of the studied water/wastewater, depending on the expected amount of hydrocarbons in the sample. Transfer it to a separatory funnel. Pipette 25 mL of tetrachlorethylene and also add it to the separatory funnel. Shake vigorously for at least 3 minutes. Drain approximately 20 mL of the tetrachloroetylene layer into the conical flask equipped with a glass joint. Add approx. 2 g of SiO_2 or Al_2O_3, stir and leave for 20 minutes, shaking a few times. Drain the rest of the organic solution to the container with C_2Cl_4. After 20 minutes filter the mixture of tetrachlorethylene solution and sorbent. Fill the measuring cell with the clear filtrate. Record the spectrum in the range between 2500 and 3500 cm^{-1}.

REPORT

1. Using the linear regression equation calculate the concentration of mineral oil in the extractant in mg/mL in the studied samples. Next, knowing the volume of water taken for analysis, determine the concentration in mg/L. Collect the results in the table below:

No. of water sample	volume of water / wastewater	A (2926 cm^{-1})	The concentration of the mineral oil in the extractant [mg/mL]	The concentration of the mineral oil in the water / wastewater [mg/L]

Compare the obtained results with the relevant regulations and draw conclusions concerning both the degree of contamination of the studied water/wastewater and applicability of this measurement method.

Bibliography

1. Szczepaniec-Cięciak, E.; Kościelniak, P.; (ed.), Environmental Chemistry, Vol. 1, 565-572, Kraków 1999. *in Polish*
2. Jakubowski M.; *n-Hexane. Documentation.* Principles and Methods of Assessing the Working Environment, 1(47), 109-129. *in Polish*
3. Bieg, B .; Kirkiewicz, J.; Krogulec, M .; *The application of the phenomenon of total internal reflection in the study of water pollution with petroleum products,* Scientific Papers of the Szczecin Maritime University 1(73), 41-48. *in Polish*
4. Pigoń K Róziewicz Z .; Physical Chemistry, v.1, PWN, Warsaw 2005.
5. Polish Standard PN-82 C-04565.01 *"Determination of the content of crude oil and its components. Determination of non-polar aliphatic hydrocarbons with infrared spectroscopy"*

6.5. Determination of absolute configuration using vibrational circular dichroism

Piotr F. J. Lipiński

A note to the instructor: This is a solely computational exercise. Nevertheless, it is more than likely that you will be unable to run all the required simulations during the class. Depending on your hardware or access to supercomputing resources, it may take even a few days. I would recommend that the Instructor have all the simulations done prior to the class and provide the students with the outputs after they set on their runs.

Many students might not have met computational chemistry methods and software before. It would be wise to use this class as a hands-on introduction to the practice of quantum-chemical calculations, especially in respect to getting to know how to use the software.

6.5.1. Determination of absolute configuration via VCD

If you want to determine absolute configuration of a compound in your sample via Vibrational Circular Dichroism, you will not manage without quantum-chemical calculations. VCD experimental measurements are usually accompanied by a more or less intensive computational modelling. This is because there are no empirical rules that would ascribe band signs (or other spectral parameters) to particular enantiomers.

Quantum-chemical packages, like Gaussian [1], can simulate VCD spectra using the theory devised by Stephens et al. [2]. VCD intensity, or more strictly rotatory strength, is calculated by taking derivatives of the energy of the molecule with respect to electric field or magnetic field and vibrational mode displacement or momentum.

VCD spectra are most often calculated with DFT methods. The applied functionals are above all B3LYP and B3PW91. As to the basis set, 6-31G(d) seems an absolute minimum for purposes of absolute configuration assignments. [3] When simulating hydrogen bonded systems, it may be reasonable to include additional diffuse or polarization functions. In general, larger basis sets improve the results, but at the expense of the computational time. [2]

In order to assign absolute configuration to the molecules in the sample, the obtained computational and experimental spectra are compared. In an ideal situation all bands observed in the experimental spectrum can be correlated to the simulated spectrum with respect to the frequencies (usually scaled to correct for the anharmonicity effects), signs and relative intensities. If the spectra correspond directly, the sample molecule is assigned with the configuration of the enantiomer for which the simulations were done. If the correspondence is inverse, the assignment is opposite.

If the studied molecule is flexible, one needs to characterize its conformers. To this aim, the conformational space is searched for low-energy conformations. Usually, conformers lying energetically 2.0 kcal/mol over the global minimum do not contribute much to spectra. However, even if those above this threshold are excluded, it still may be sometimes time-consuming (but most often indispensable) to simulate spectra for all the low-energy conformations. Boltzmann-population-weighted composite theoretical spectra may be then compared to the experimental results. It is however worth to note that the weighting factors change with the theoretical method, and are prone to error. In some cases, one can find a band (or a few of them) that do not change much (especially as to their sign) with the conformation – in such cases it can be a diagnostic band for the configuration assignment.

Finally, simulating the spectra, one has to be aware of possibility of effects such as molecular association (e.g. dimerization in concentrated solutions) or strong solvent effects.

The above remarks are far from being comprehensive of the field. The art of assigning absolute configuration *via* VCD is described in more detail in Refs [4–11].

AIM OF THE EXPERIMENT

1. Determine absolute configuration of a 1-phenylethanol (Figure 6.5.1) sample. Its VCD spectra are reproduced in Figure 6.5.2.
2. Determine the prevalent conformation of the molecule in a solution.
3. Explore the effect of method and basis set on the parameters of vibrational spectra.
4. Explore the effect of a substituent on the parameters of vibrational spectra.

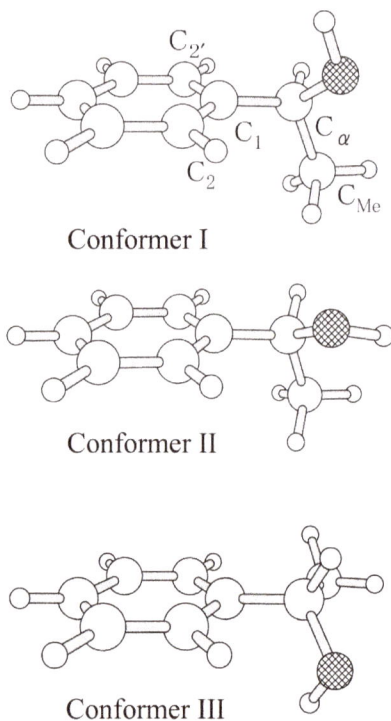

Conformer I

Conformer II

Fig. 6.5.1. Conformers of 1-phenylethanol.
Reprinted with permission from Ref. [12]
Copyright 2007 American Chemical Society.

Conformer III

SCIENTIFIC BACKGROUND

1. Understanding the need of configurational assignments.
2. Understanding fundamentals of vibrational spectroscopy.
3. Understanding fundamentals of computational chemistry.
4. Basic computer literacy.

EQUIPMENT, MATERIALS

1. A modern PC with Gaussian and GaussView. [1]

PROCEDURE

1. We shall study 1-phenylethanol (1-PhEtOH) – an important intermediate in the synthesis of chiral compounds. Its chirality results from the presence of an asymmetric carbon. 1-PhEtOH is a flexible molecule. It has been previously determined [12] *via* conformational analysis that it has three stable conformers in the gas-phase (at the B3LYP/6-311++G(d,p) level) with the following values of C2-C1-Cα-O and C1-Cα-O-H dihedral angles:

I. -43°, -55°   II. -29°, 173°   III. 47°, 57°

Prepare structures of these conformers for (R)-1-phenylethanol in GaussView or other molecular graphics software.

2. Optimize the structures and calculate their VCD spectra at the B3LYP/6-311+ +G(d,p) level.

3. Additionally, optimize the structure and calculate the VCD spectrum of the conformer I at the following levels:
   a) HF/6-31G
   b) B3LYP/6-31G
   c) B3LYP/6-311G
   d) B3LYP/6-31G(d,p)
   e) B3LYP/6-311G(d,p)
   f) B3LYP/6-31+G(d,p)
   g) B3LYP/6-31++G(d,p)
   h) B3LYP/6-311+G(d,p)
   i) B3LYP/6-311++G(d,p)
4. Take the optimized structure for the conformer I and prepare inputs for the *para*-derivatives with the following substituents: -BF_2, -Cl, -F, -H, -OH (two conformers), -NH_2. Optimize them and calculate VCD spectra at the HF/STO-3G level. (This level is normally inadequate for studying VCD spectra; however, you will use it for the sake of calculation speed. We checked that it gives qualitatively the same results for these molecules as some more computationally demanding methods).

REPORT

1. Compare the simulated spectra for each conformer of with the fingerprint region of an experimental VCD spectrum of one of the enantiomers of 1-PhEtOH acquired in a CS_2 solution at 0.1 M. (Figure 2a)
   Determine the absolute configuration of the enantiomer in the measured sample. Are you able to say which of the conformer is prevalent in the CS_2 solution? If so, which one? Is this prevalence in line with the conformational preferences for the gas-phase?

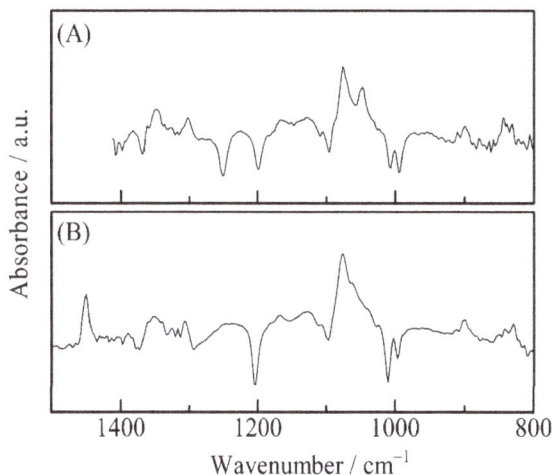

Fig. 6.5.2. VCD spectra of 1-phenylethanol acquired in a) CS_2 solution b) liquid state. Reprinted with permission from Ref. [12] Copyright 2007 American Chemical Society

2. 1-PhEtOH spectrum acquired in the liquid state (Figure 2b) is slightly different from that obtained in the CS_2 solution (Figure 2a). Why?

3. From the outputs of the Point 2. of exercise procedure, take one of the well-identifiable normal modes over 2700 cm^{-1} (e.g. O-H stretching) and compare its frequency, infrared intensity and rotatory strength across the various methods. Do the same for one of the modes in the fingerprint region. What changes? (compare values and signs). Are the modes computationally stable? Report also the calculation times.

4. For the simulated derivatives, plot the wavenumbers, infrared frequencies and rotatory strengths (VCD intensities) of O-H stretching and methane C-H stretching against Hammett σ_p substituent constants [13] and pEDA constants (Table 1). [14] Which of the values are correlated? Why or why not? (consider conformation and the meaning of the descriptors). Report the correlation coefficients. If you are interested in substituent effects on VCD spectra, you might read Ref [15, 16].

Table 6.5.1. Values of substituent constants

Substituent	σ_p [13]	pEDA [14]
BF$_2$	0.48	-0.078
Cl	0.23	0.064
F	0.06	0.068
H	0.00	0.000
OH	-0.37	0.114
NH$_2$	-0.66	0.141

References

1. Frisch M.J., et. al., *Gaussian 09, Revision B.01*, (2009).
2. Stephens P.J., Devlin F.J., *Determination of the structure of chiral molecules using ab initio vibrational circular dichroism spectroscopy.*, Chirality, **12**, 172 (2000).
3. Kuppens T., Langenaeker W., Tollenaere J.P., Bultinck P., *Determination of the Stereochemistry of 3-Hydroxymethyl-2,3-dihydro-[1,4]dioxino[2,3- b]- pyridine by Vibrational Circular Dichroism and the Effect of DFT Integration Grids*, J. Phys. Chem. A, **107**, 542 (2003).
4. Stephens P.J., Devlin F.J., Pan J.-J., *The determination of the absolute configurations of chiral molecules using vibrational circular dichroism (VCD) spectroscopy.*, Chirality, **20**, 643 (2008).
5. Shah R.D., Nafie L.A., *Spectroscopic methods for determining enantiomeric purity and absolute configuration in chiral pharmaceutical molecules.*, Curr. Opin. Drug Discov. Devel., **4**, 764 (2001).
6. Kuppens T., Bultinck P., Langenaeker W., *Determination of absolute configuration via vibrational circular dichroism*, Drug Discov. Today Technol., **1**, 269 (2004).
7. McConnell O., He Y., Nogle L., Sarkahian A., *Application of chiral technology in a pharmaceutical company. Enantiomeric separation and spectroscopic studies of key asymmetric intermediates using a combination of techniques. Phenylglycidols.*, Chirality, **19**, 716 (2007).

8. Mcconnell O., Ii A.B., Balibar C., Byrne N., Cai Y., Carter G.U.Y., Chlenov M., Di L.I., Fan K., Goljer I., He Y., Herold D.O.N., Kagan M., Kerns E., Koehn F., Kraml C., Marathias V., Marquez B., Mcdonald L., Nogle L., Petucci C., Schlingmann G., Tawa G., Tischler M., Williamson R.T., Sutherland A., Watts W., Young M., Zhang M., Zhang Y., Zhou D., Ho D., *Review Article Enantiomeric Separation and Determination of Absolute Stereochemistry of Asymmetric Molecules in Drug Discovery — Building Chiral Technology Toolboxes*, **682**, 658 (2007).

9. Polavarapu P.L., *Renaissance in chiroptical spectroscopic methods for molecular structure determination.*, Chem. Rec., **7**, 125 (2007).

10. Freedman T.B., Cao X., Dukor R.K., Nafie L.A., *Absolute configuration determination of chiral molecules in the solution state using vibrational circular dichroism.*, Chirality, **15**, 743 (2003).

11. Polavarapu P.L., *Molecular structure determination using chiroptical spectroscopy: where we may go wrong?*, Chirality, **24**, 909 (2012).

12. Shin-ya K., Sugeta H., Shin S., Hamada Y., Katsumoto Y., Ohno K., *Absolute configuration and conformation analysis of 1-phenylethanol by matrix-isolation infrared and vibrational circular dichroism spectroscopy combined with density functional theory calculation.*, J. Phys. Chem. A, **111**, 8598 (2007).

13. Hansch C., Leo A., Taft R.W., *A survey of Hammett substituent constants and resonance and field parameters*, Chem. Rev., **91**, 165 (1991).

14. Oziminski W.P., Dobrowolski J.Cz., σ – *and* π-*electron contributions to the substituent effect: natural population analysis*, J. Phys. Org. Chem., **22**, 769 (2009).

15. Lipiński P.F.J., Dobrowolski J.Cz., *Substituent effect in theoretical VCD spectra*, RSC Adv., **4**, 27062 (2014).

16. Lipiński P.F.J., Dobrowolski J.Cz., *Local chirality measures in QSPR: IR and VCD spectroscopy.*, RSC Adv., **4**, 47047 (2014).

6.6. The identification of painting materials and degradation products. FT-IR imaging of paint layers

Emilia Staniszewska-Ślęzak, Kamilla Malek

Over the last few years there has been an increasing interest in using infrared imaging technique to analyse a variety of works of art. FT-IR microscopy enables the non-destructive identification of painting materials from samples with a size of several microns. The results obtained, in particular by infrared imaging, also show the spatial distribution of the identified substance. Information gathered in such a way is very helpful for conservators in restauration of works of art and they are used to describe the artist's workshop.

6.6.1. Chemical characteristics of a paint layer and products of its degradation

Firstly, one should consider what exactly an object of this research is. Conservators mainly work on cross-sections through all the layers of painting taken from works of art. This sampling strategy illustrates precisely the stratigraphy of the studied layers of paint. For this purpose, a sample taken most often in the conservation process is embedded in a block of a resin, such acrylic polymers, polyester and epoxy. Then the samples are cut in order to obtain cross-sections and next it is polishing to obtain an even, flat surface to be analysed.

Figure 6.6.1 shows an exemplary cross section through the paint layer prepared for collection spectra in infrared imaging while a microphotography of a sample exhibits a size of a cross-section and the individual paint layers. A paint layer is a mixture of many substances of the organic and inorganic origin, which includes a group of coloured substances (pigments, dyes), fillers, and binders. Substances formed due to the action of environmental factors such as high humidity, abnormal temperature, UV -Vis radiation or impurities are degradation products. A characteristic of coloured substances and fillers is presented in Chapter 7.7 whereas binders and their degradation products are described below.

Binders are liquid or semi-liquid substances, in which grains of pigments and fillers are suspended. Their role is to keep these substances within the layer and to attach a paint layer to a painting ground support [1,2]. Taking into account their chemical origin binders are divided into oils, waxes, resins, synthetic adhesives, balsams, protein binders, and sugar and water-emulsion temperas [2]. The most common binding media are oil and proteins-based binders.

Oil binders belong to three classes of drying, semi-drying and non-drying oils [1]. Drying oils are glycerol esters of unsaturated, *i.e.* linoleic, linolenic and oleic acids, and saturated fatty acids, *i.e.* stearic and palmitic acids. Examples of such oils are oils from linseed, poppy seed, walnut, hemp and Chinese oil. They dry out in the air due to the oxidation process of unsaturated fatty acids and produce linoxyn, which forms thin and transparent film resistant to environmental conditions. Some metal ions and their compounds from pigments catalyse the process of oils drying, e.g. lead or cobalt compounds [2]. The most widely used oil binder is linseed oil or semi-drying sunflower and soybean oils [1]. Oils as triglycerides containing

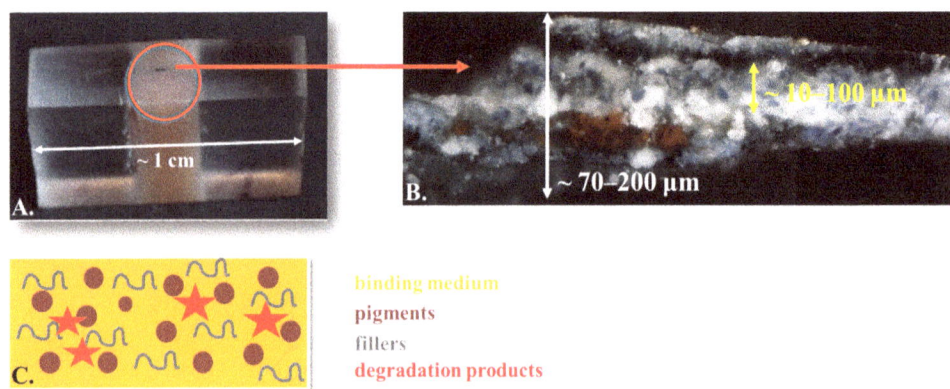

Fig. 6.6.1. A paint chip embedded in a resin (A.) a microphotography of a cross-section (B.) a schematic showing composition of a paint layer (C.)

multiple C=C bond have characteristic IR bands at 1740 and 2850–2930 and 3010 cm[-1] (a detailed description of IR spectra of lipids is provided in Chapter 6.8). Unfortunately FT-IR technique does not identify the type of oil.

The protein-based binding media obviously contain proteins that are polymers of amino acids, in which amino acid residues are bonded by peptide bonds. This type of binders is known since ancient times [2] and in order to prepare them egg white, egg yolks or whole eggs, milk, casein, skin glue were mainly used. Since medieval times decoction of dried swim bladders of sturgeon was also used [2]. Mixing the protein adhesive with water gives tempera, otherwise known as emulsion, which is another type of binding medium [1]. Tempers are divided into some classes according to their composition, e.g. rubber, starch, rye flour emulsions [2]. Similarly to oil binders, protein-based medium is easily detected by FT-IR spectroscopy by the presence of bands typical for proteins (Chapter 6.9). Particularly, amide A, I and II bands at 3200, 1640 and 1540 cm[-1], respectively, are the marker bands, although the amide II band is often overlapped by pigments bands.

So far a process responsible for aging of paint layers containing drying oils has not been unambiguously explained. Some mechanisms are based on radical reactions involving fatty acids, which begin from a removal of the hydrogen atom from the methyl group between the two C=C bonds. Then, the radical may be bonded by the adjacent double C=C bond of fatty acid to form a molecule with a higher molecular weight than the parent fatty acid or be conjugated to the double C=C bond within the same fatty acid yielding to intramolecular cyclization. However, these reactions are slow and take place only in heating of oil in anaerobic conditions. Another pathway is the formation of the C-C bonds in the presence of oxygen or a reaction with a radical thereby forming hydroperoxides.

The latter are extremely reactive and very quickly undergo a series of transformations such as polymerization or termination. An important step in the process of the binder drying is the degradation of fatty acids to form aldehydes and alcohols with shorter chains. This process is observed by a band-shift of the stretching vibration of the C=O bond (at 1740 cm[-1] for triglycerides) towards lower wavenumbers (up to1700 cm[-1]).

Next, these compounds in the presence of oxygen are oxidized to the corresponding carboxylic acids which can react with certain metal ions of pigments producing soaps – salts of fatty acids. These salts are responsible for the degradation process of paint layers causing the destruction of works of art. So far copper (II), zinc and lead (II) stearates, palmitates and oleates have been identified along with zinc linoleate, lead (II) azelate calcium and copper (II) oxalates [3-5]. Bands of the soaps appear in the range of 1500 – 1600 cm[-1] of IR spectra and usually are not overlapped by bands of pigments, fillers and binders therefore their presence suggests the degradation of paint layers. Marker bands listed-above metal salts were determined from their reference spectra. An obstacle in the IR analysis of these degradation products is the lack of a detailed description of the various stages of the binder oxidation process and the possibility of the formation of many chemicals. Their unambiguous identification requires an application of complementary analytical techniques, e.g. elemental analysis and mass spectrometry.

6.6.2. FT-IR spectroscopy as a technique used in the identification of chemical composition of artworks

FT-IR spectra show marker bands characteristic for all of the components listed-above assuming that vibrations of a substance are active in the spectral range available on a FT -IR spectrometer/microscope. It depends on an employed detector (see Section 1.3). In the case of imaging systems one can collect IR spectrum is the range above 850 cm^{-1}. Thus the majority of pigments consisting of simple salts (e.g. minium, vermilion, titanium white) is not identified in mid IR spectra but their presence is revealed by Raman spectroscopy (see Chapter 7.7). As is well-known given functional groups have IR bands in a specific spectral region, thus the main marker bands for example for carbonate ions in chalk (calcium carbonate) and white lead (lead carbonate) are present in the same spectral range. In this case, the identification of a particular substance must be confirmed by other bands. For instance, a band at 780 cm^{-1} is characteristic for chalk only. FT-IR imaging spectroscopy, for the purposes presented in this chapter, can be employed by using two techniques: reflection (if a studied object exhibits a high reflectance

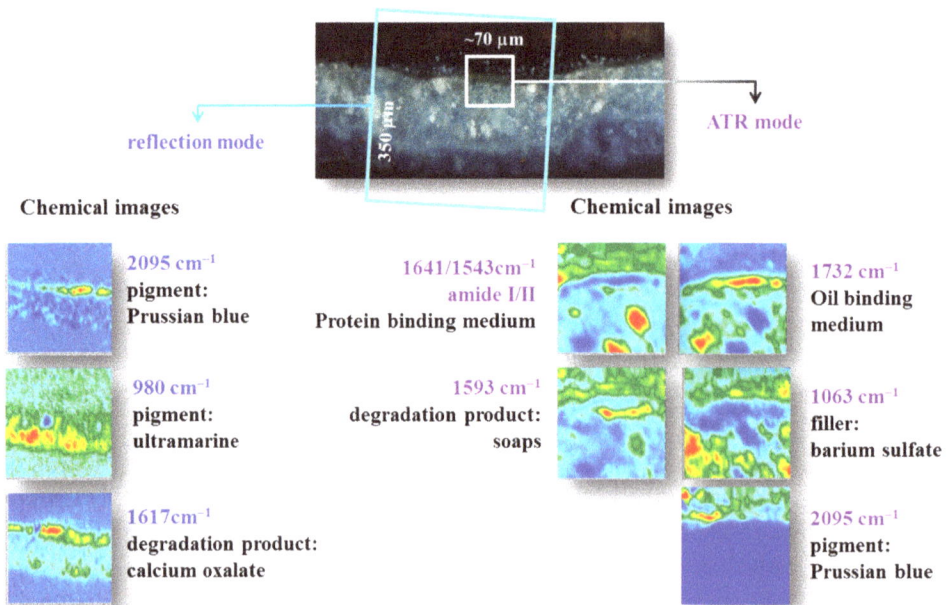

Fig. 6.6.2. A comparison of FT-IR imaging results for a cross-section of paint layers performed by using reflection and ATR modes

index) and ATR (attenuated total reflectance). Fundamentals of infrared imaging are described in Chapter 3.2.

Figure 6.6.2 shows imaging of both techniques. The advantage of imaging in reflection is one measurement of all layers at the same time as this mode allows to record spectra from the area of approx. 350×350 µm^2. However, due to weak reflectance of painting materials in a layer this technique reveals the presence a few components only (only three substances were detected in a cross-section shown in Fig. 6.6.2) and provides chemical images with maximum spatial resolution of 5-6 microns only. A better spatial resolution and sensitivity shows ATR imaging mode but it requires several measurements to cover the entire area of a cross-section, see Fig. 6.2.2. An analysis and construction of chemical images illustrating spatial distribution of a component are performed by two methods; selecting a marker band and calculating its integral intensity as shown in Fig. 6.6.3, or by using a chemometric analysis (mainly cluster analysis and principal component analysis, *see* Chapter 5).

The identification of individual substances is mainly based on a comparison with spectra of reference substances. Chemometric analysis is extremely effective if chemical components present in a layer are chemically similar and integration of overlapping marker bands can lead to false conclusions [6,7]. This problem is illustrated in Figure 6.6.3, where gypsum and barium sulphate in the form

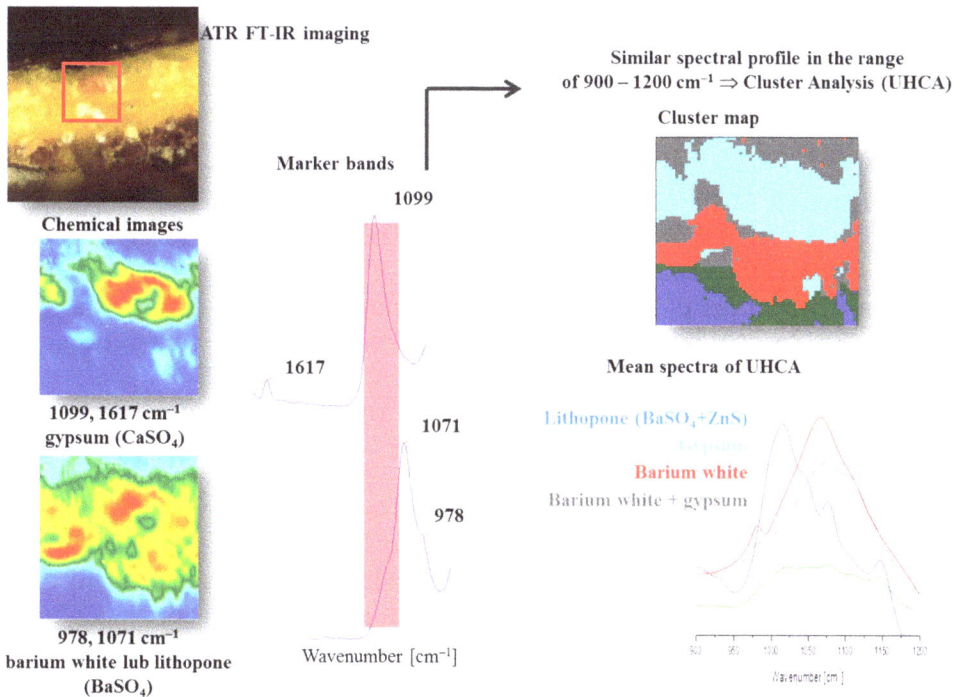

Fig. 6.6.3. An analysis of ATR FT-IR imaging results by bands integration (labelled as chemical images) and cluster analysis (UHCA, unsupervised hierarchical cluster analysis)

of lithopone or white barite were detected. Their marker bands appear at similar wavenumbers generating false images of their distribution. While cluster analysis detected the type of sulphate salts as well as determined the distribution and co-existence of found chemicals in the paint layer.

AIM OF THE EXPERIMENT

1. Learning measurements of ATR FT-IR imaging of paint layers in the form of a cross-section.
2. Interpretation of IR imaging results by the integration and chemometric analysis.

SCIENTIFIC BACKGROUND

1. FT-IR imaging and microscopy; Chapter 3.3.
2. Chemometric analysis; Chapter 5.
3. Chemical composition of a paint layer and its degradation (additional information is given in Chapter 7.7).

FURTHER READING

1. E. Staniszewska, K. Malek, Z. Kaszowska, *Studies on paint cross sections of a glass painting by using FT-IR and Raman microspectroscopy supported by univariate and Hierarchical Cluster analysis*, J. Raman Spectrosc., **44**, 1144 (2014).

EQUIPMENT, MATERIALS, CHEMICALS

1. Cross sections of paint layers provided by an assistant.
2. IR microscope with a ATR crystal (e.g. germanium).
3. Software for spectral analysis, e.g. Opus, CytoSpec.

PROCEDURE

1. Perform ATR FT-IR imaging of a cross-section of paint layers.
2. Construct chemical images for bands present in collected spectra showing spatial distribution of substances.
3. Identify the origin of bands based on the literature [1,6,7] and databases of reference spectra: http://tera.chem.ut.ee/IR_spectra/ ; http://www.irug.org/search-spectral-database.
4. Perform hierarchical cluster analysis of the imaged area of a sample.

REPORT

1. Identify painting materials present in a cross-section based on FT-IR spectra of reference substances and classify them into respective groups of painting materials, *i.e.* pigments, binders, fillers or degradation products. Comment their appearance correlating chemical images with paint layers visible in a photomicrography.
2. On the basis of maps derived from cluster analysis and their mean spectra, discuss the distribution and co-occurrence of the identified substances. Comment compatibility of chemical images with cluster maps.

3. Discuss the utility of FT-IR microscopy to study cross-sections of the paint layers.

References

1. Vahur S., *Expanding the possibilities of ATR FT-IR spectroscopy in determination of inorganic pigments*, PhD thesis, Tartu University, Estonia, 2010 (available at http://dspace.utlib.ee/dspace/bitstream/10062/14740/1/vahur_signe.pdf).
2. Hopliński J., *Farby i spoiwa malarskie*, Publishing house: Zakład Narodowy im. Ossolińskich, Wyd. II, Wrocław 1990 (in Polish).
3. Robinet L., Corbeil M.C., *The characterization of metal soaps*, Studies Conserv., **48**, 23 (2003).
4. Plater M.J., De Silva B., Gelbrich T., Hursthouse M.B., Higgitt C.L., Saunders D.R., *The characterisation of lead fatty acid soaps in 'protrusions' in aged traditional oil paint*, Polyhedron, **22**, 3131 (2003).
5. Nevin A., Loring Melia J., Osticioli J., Gautier G., Rerla Colombini M., *The identification of copper oxalates in a 16th century Cypriot exterior wall painting using micro FTIR, micro Raman spectroscopy and Gas Chromatography-Mass Spectrometry*, J. Cult. Her., **9**, 154 (2008).
6. E. Staniszewska, K. Malek, Z. Kaszowska, *Studies on paint cross sections of a glass painting by using FT-IR and Raman microspectroscopy supported by univariate and Hierarchical Cluster analysis*, J. Raman Spectrosc., **44**, 1144 (2014).
7. Kaszowska Z., Malek K., Panczyk A., Mikolajska A., *A joint application of ATR-FTIR and SEM imaging with high spatial resolution: Identification and distribution of painting materials and their degradation products in paint cross sections*, Vib. Spectrosc., **65**, 1 (2013).

6.7. Structural analysis of proteins by means of FT-IR spectroscopy

Katarzyna Majzner

6.7.1. Anatomy of proteins

Describing molecular structure of proteins in a hierarchical way, four types of structure need to be considered:

1) the primary structure is the composition of amino acid residues and their linear sequence in the polypeptide chain;
2) the secondary structure is in turn spatial arrangement of the peptide backbone resulting from interactions between the amide bonds stabilized *via* hydrogen bonding. This structure is defined by three torsion angles illustrated in Figure 6.7.1;
3) the tertiary structure expresses spatial arrangement of the whole polypeptide chain formed due to interactions between amino acid residues *R*, e.g. the formation of disulphide bridges, van der Waals forces;

4) the quaternary structure presents spatial arrangement of at least two polypeptide chains (subunits of protein) including possible contribution of non-pro-
   tein structures e.g. heme.

The determination of the spatial structure of proteins, especially their folding
in a native state is strongly correlated with their function in living organisms. The
spatial arrangement of the polypeptide chain is also called conformation of protein.

In turn, vibrational spectroscopy is one of the techniques that has been widely
used in studies on secondary structure of proteins. An application of Raman spectroscopy in the analysis of protein structure is described in Chapter 7.1. As mentioned above the parameters used in the determination of the secondary structure
are the three torsion angles around the peptide bond and repeated regularly along
the peptide backbone. Figure 6.7.1 depicts these angles that are defined as follows
ω (C_α-C(O)-N-C_α), ψ (-N-C_α-C-N-) and φ (-C-N-C_α-C-).

The torsion angle ω is always $\sim 180°$, which means that the atoms involved in
its formation lie in one plane. Planarity of the peptide bond results from its chemical nature since this is a partially double bond formed between the carbonyl carbon
atom and the imino nitrogen atom. This leads to the coexistence of two mesome-
ric/resonance forms that quite effectively restrict rotation around the -C(O)-N-
bond (Figure 6.7.2). Mesomeric forms are a consequence of overlapping p orbitals
of the nitrogen and carbon atoms [1].

Since the ω angle is always planar, the protein conformational analysis relies,
in fact, on the two dihedral angles φ and ψ. Rotation around the -C_α-C- and -N-C_α-
bonds is relatively free and the accepted values of the φ and ψ angles determine
the type of secondary structures. If secondary structure is considered as locally organized fragments of the polypeptide chain with some values of φ and ψ angles,
two main types of secondary structures are defined as α-helices and β-sheets. Note
that due to steric hindrance, the angles φ and ψ cannot accept any value and this
fact was firstly observed by G.N. Ramachandran. Apart from α-helix and β-sheet

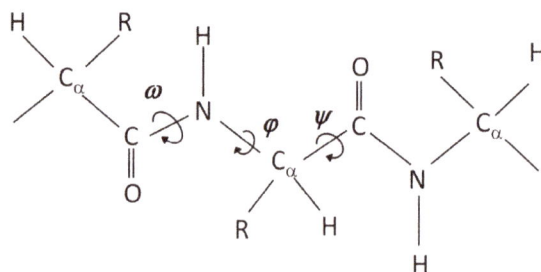

Fig. 6.7.1. Scheme of the peptide
bond with marked torsion angles

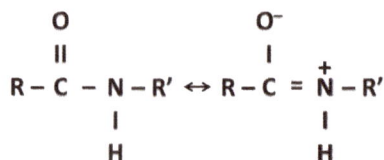

Fig. 6.7.2. Mesomeric forms of the peptide bond

structures, a few other conformations are found in nature. For example, 3-helix is found in collagen. This conformation is composed of three polypeptide chains with a specific sequence of three amino acid residues, *i.e.* Gly-Pro-Hyp. This form differs from a typical helical conformation since H-bonding is not form within a single chain but results from interactions between adjacent amino acids residues present in separated three polypeptide backbone. Such a structure is called superhelical structure. Stabilization forces in superhelical structure are formed due to repulsion of the pyrrolidinyl rings in proline and hydroxyproline. Other secondary structures result from interactions between amino acid residues in a fragment of the peptide sequence like in β-turn conformation or from interactions between α-helix and β-sheet structures. Such structures are called in general structural motifs, e.g. β-hairpin consists of two antiparallel beta strands connected by a tight turn of a few amino acids or "Greek key" is formed from folding of four β strands into a sandwich shape. In addition, it is well known that within regions composed of several secondary conformations there are fragments in the peptide backbone that show unordered structure. Unordered structures play an important biological role as they can change conformation into an ordered one due to biological processes such as ligand binding.

The most common structure in nature is α-helix conformation. α-helix structure has been identified by Pauling and Corey in α-keratin present in hair and wool [2]. Amino acid residues are located outwardly from the center of the helix forming a polar coat while the peptide bonds are directed towards the center of the non-polar cylinder.

β-sheets consist of β-strands connected laterally by at least two or three backbone *via* hydrogen bonding, forming generally a twisted, pleated sheet. This is not the only difference in structure of α-helix and β-sheet. The formation of β-sheets requires involving of a few polypeptide chains whereas α-helices are formed by one peptide backbone. In addition, the arrangement of polypeptide chains in the same direction gives parallel β-sheet conformation whereas the opposite arrangement is called an anti-parallel form.

6.7.2. Application of FT-IR spectroscopy to studying proteins

IR spectrum exhibits the presence of five protein bands called amide bands since they are attributed to vibrations of the amide bond. It is worth noticing that even though one expects the presence of bands of vibrations originating from amino acid residues they do not provide a high intensity bands in IR. Characteristic bands of proteins observed in IR spectrum are called amide I, II, III, A and B. Among them, amide I and II dominate in a spectrum and their motions are illustrated in Figure 6.7.3.

In detail, since a few modes can contribute to IR bands, amide I band represents coupling of the stretching mode of the carbonyl group –C=O ($[\nu(C=O)]$; 80%) with minor contributions from the C-N stretching vibration $[\nu(C-N)]$ and the NH in-plane

(I)   (II)

$$\underset{\text{(I)}}{\overset{\uparrow}{\underset{\downarrow}{O}}} \quad\quad \overset{O}{\underset{}{}}$$

Fig. 6.7.3. Scheme of amide I ((I), the stretch of the C=O) and amide II vibrations ((II) the in-plane bend of the NH group)

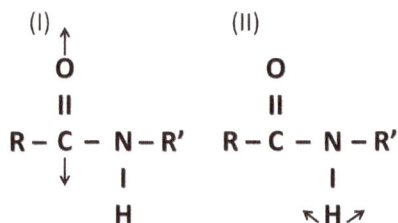

$$\begin{array}{cc} O & O \\ \| & \| \\ R-C-N-R' & R-C-N-R' \\ \downarrow \ \ | & \ \ \ | \\ H & \nwarrow H \nearrow \end{array}$$

bend [δ(N-H)]. In turn, an amide II band originates from the in-plane bending vibration of the N-H bonds ([δ(N-H)]; 60%) and 40% of the C-N stretching vibration [v(C-N)]. Amide III band is associated with 40% of the C-N stretching mode, 30% of the in-plane NH bending mode [δ(N-H)] and 20% of C-C stretching vibrations in CH$_3$–C [v(CH$_3$–C)] [11]. Table 6.7.1 and Figure 6.7.4 summarize spectral features of the amide modes.

Amide A band is present in IR spectrum in the region above 3000 cm^{-1} and is assigned to the Fermi resonance between the NH stretches and the first overtone of an amide II band [3]. Its maximum is also sensitive to the strength of hydrogen bonding. In turn, amide B band is assigned to the stretching vibration of intramolecular hydrogen bonds between the NH groups [3]. The other amide bands are summarized in Table 2 and they are not usually use in the analysis of the secondary structure of proteins from IR spectra.

Since protein spatial structure can result from a number of different secondary conformations, infrared spectra of proteins are complex. Observed amide bands

Table 6.7.1. Spectral features of amide bands in IR spectrum, including the assignment of the dominant vibrations [8]

Amide band	Spectral ranges [cm^{-1}]	Assignment of the dominant vibrations*
I	1690–1600	70-80% v(C=O), 10-20% v(C-N), δ(N-H),
II	1580-1480	40-60% δ(N-H) 20-40% v(C-N)
III	1300-1220	40% v(C-N), 30% δ(N-H), 20% v(CH$_3$-C)
A	3300-3250	v(N-H)
B	3100-3030	v(N-H)

* v – stretching vibration, δ – in-plane bending vibration,

Table 6.7.2. Spectral ranges of amide IV-VII bands in IR spectrum along with their tentative assignment [8,9]

Amide band	Spectral ranges [cm^{-1}]	Assignment of the dominant vibrations*
IV	770-625	δ(OCN)
V	800-640	γ(N-H)
VI	610-540	γ(C=O)
VII	200	backbone vibrations

* δ – in-plane bending vibration, γ – out-of-plane bending vibration

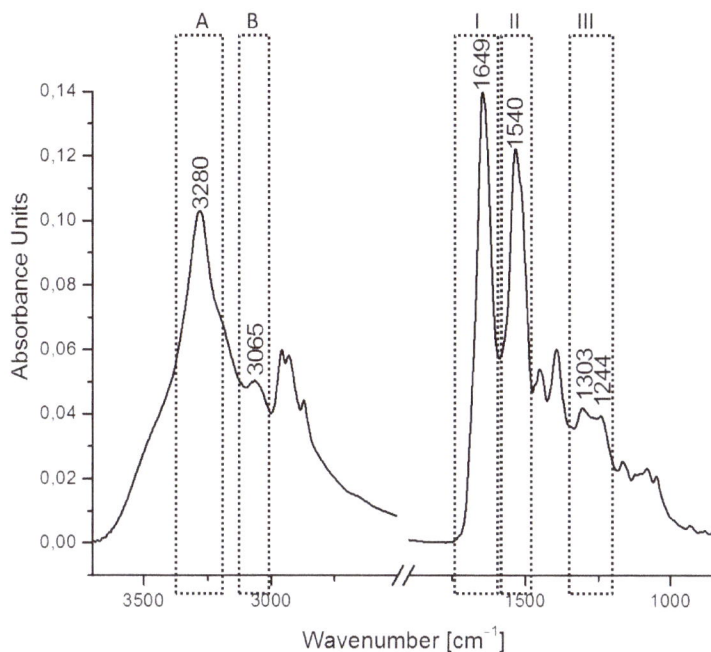

Fig. 6.7.4. An exemplary ATR FT-IR spectrum of a protein (trypsin inhibitor) with labelled amide bands

consist of overlapping bands specific for particular structures. The most important spectral marker for secondary structures is the position of amide I band. Table 6.7.3 shows spectral ranges typical for each conformation of proteins.

In order to obtain detailed information on components of IR bands, which can then be assigned to a specific secondary structure, several procedures can be applied. For instance, we can use deconvolution of bands, deconvolution using inverse Fourier transformation, and calculation of second derivative spectra. The last method allows for a relatively quick indication of bands positions without a need of artificial generation of components like in the case of bands deconvolution. In such a spectrum, bands are represented as minima. Figure 6.7.5 shows an example of second derivative spectrum of protein with labelled positions of the observed bands. The shape of the second derivative depends on quality of recorded spectra. The second derivative spectra are very sensitive to any disturbance (e.g. noise

Table 6.7.3. Spectral ranges of amide I bands characteristic for secondary structures [3]

Secondary structure	Spectral range of amide I band [cm^{-1}]*
α-helix	1640 -1660 (s)
β-sheet parallel	1625 – 1640 (s)
β-sheet antiparallel	doublet 1620 (s), 1680 – 1690 (w)
unordered	1640 – 1660 (s)
β-hairpin, loop	1660 – 1680 (s)

*s – strong; w – weak

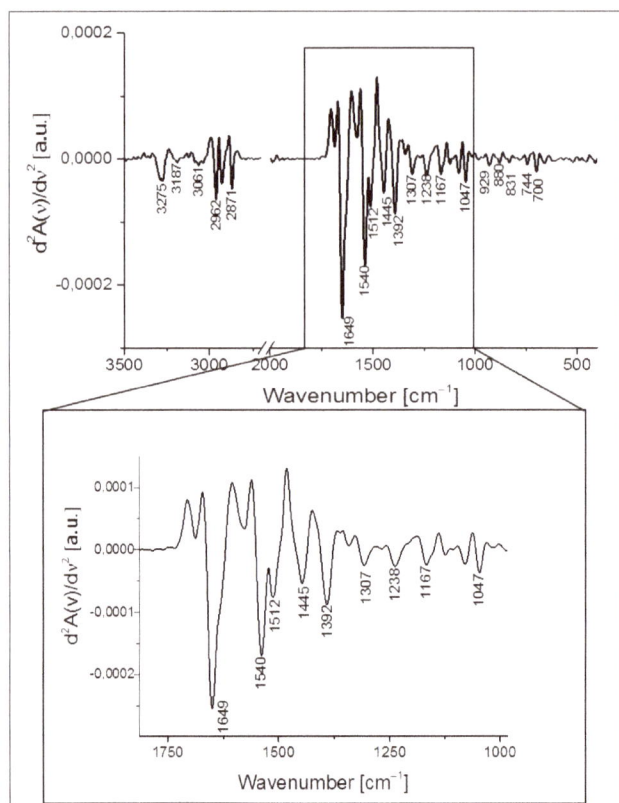

Fig. 6.7.5. Second derivative ATR FT-IR spectrum of protein (trypsin inhibitor) with labelled positions of observed bands , where A(v) – wavenumber-dependent absorbance, v – wavenumber

of the instrument, water vapour uncorrected absorption bands), and therefore its shape should be carefully analysed.

AIM OF THE EXPERIMENT

1. The analysis of proteins by infrared absorption spectroscopy;
2. Correlation of protein secondary structure and marker bands in the FT-IR spectrum;
3. Interpretation of FT-IR spectra of samples with a biological origin.

SCIENTIFIC BACKGROUND

1. Fundamentals of absorption FT-IR spectroscopy; Chapter 1.
2. Construction and operation of FT-IR spectrometer, including the ATR technique; Chapter 3.1.
3. Basics of the protein structure.

EQUIPMENT, MATERIALS, CHEMICALS

1. Solid protein standards: albumin, hemoglobin, trypsin, α-chymotrypsinogen A, papain, α-lactalbumin, a protease, a sample of biological origin containing a defined type of protein brought to class by the student.

2. Fourier transform IR spectrometer equipped in ATR crystal (diamond), e.g. Bruker Alpha.
3. Software capable to a graphical spectrum presentation and calculation of the second derivative, e.g. OPUS software.

PROCEDURE

1. Measure ATR FT-IR spectra of reference substance, including optimizing measurement conditions (number of scans for background and sample) allowing to obtain adequate signal to noise ratio in the shortest possible time in the whole spectral range.
2. Determine and indicate the protein marker bands in collected spectra. Calculate the second derivative of the spectrum and use it to indicate protein marker bands of particular secondary structure of investigated reference samples (α, β, α and β mixed).
3. Measure ATR FT-IR spectrum of a biological sample.
4. Indicate positions of the marker bands like for the reference spectra in 2.

REPORT

1. On the basis of the reference spectra indicate marker bands for proteins and assign them to amide vibrations. Discuss differences and similarities in spectra of proteins with different secondary conformations (α and β).
2. Read from a spectrum of a biological origin sample positions of the amide bands, and then assign positions to secondary structures using reference spectra. Discuss an identified structure of a protein with its expected conformation reported by the literature.
3. Discuss the utility of ATR FT-IR method to study secondary structure of proteins and their identification in samples of a biological origin.

References

1. McMurry J., *Chemia Organiczna*, v. 5, WN PWN, Warszawa 2001.
2. Robert K. Murray, Daryl K. Granner, Peter A. Mayes, Victor W. Rodwell, *Biochemia Harpera*, WL PZWL, Warszawa, 1995.
3. Barth A., *Infrared spectroscopy of proteins*, BBA, **1767**, 1073 (2007).
4. Wiśniewski M., Sionkowska A., Kaczmarek A., Lazare S., Tokarev V., *Wpływ promieniowania laserowego na cienkie błony kolagenowe*, POLIMERY, **52**, 259 (2007).
5. Griffiths P. R., De Haseth J. A., *Fourier Transform Infrared Spectrometry*, John Wiley & Sons, New Jersey 2007.
6. Surewicz W. K., Mantsch H. H., Chapman D., *Determination of protein secondary structure by Fourier transform infrared spectroscopy: a critical assessment*, Biochemistry **32**, 389 (1993).
7. Barth A, Zscherp C., *What vibrations tell us about proteins*, Q Rev Biophys, **35**, 369 (2002).
8. Carbonaro M., Nucara A., *Secondary structure of food proteins by Fourier transform spectroscopy in the mid-infrared region*, Amino Acids, **38**, 679 (2010).
9. Kong J., Yu S., *Fourier Transform Infrared Spectroscopic Analysis of Protein Secondary Structures*, ABBS, **39**, 549 (2007).

10. Kęcki Z., *Podstawy spektroskopii molekularnej*, WN PWN, Warszawa 1992.
11. Szafran M., Dega-Szafran Z., *Określanie struktury związków organicznych metodami spek-troskopowymi. Tablice i ćwiczenia*, WN PWN, Warszawa 1988.
12. Sadlej J., *Spektroskopia Molekularna*, WNT Warszawa 2002.

6.8. Structural analysis of lipids by using infrared spectroscopy

Tomasz P. Wróbel

6.8.1. Characteristics and occurrence of lipids

Lipids are a large group of chemical compounds having very important functions in biological systems: energy storage, formation of lipid membranes and signal transduction.

Structurally several lipid subgroups can be distinguished, out of which the most important are fatty acids, glycerides (including triglycerides), glycerophospholipids, sphingolipids, steroids (including cholesterol) and glycolipids. From the point of view of the functioning of the human body, it is also worth mentioning cholester-ol esters, which take active part in the metabolism of lipids. Figure 6.8.1 shows

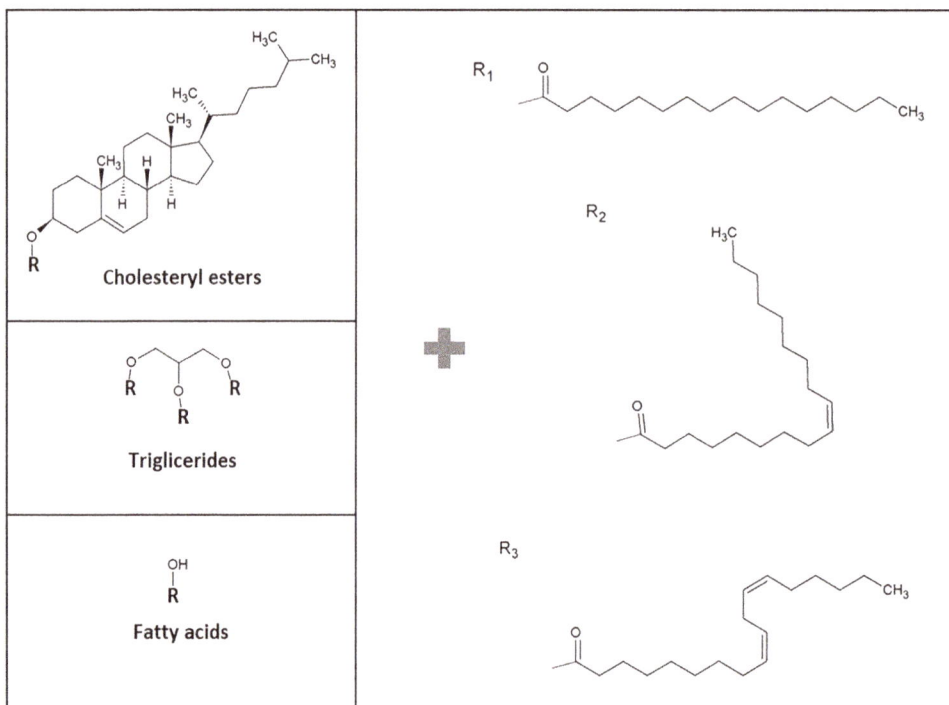

Fig. 6.8.1. Molecular structures of three groups of lipids: cholesteryl esters, triglycerides and fatty acids

simplistic structural formulae of three groups of lipids: cholesteryl esters, triglycerides and fatty acids.

6.8.2. Application of FT-IR spectroscopy to studying lipids

Infrared Spectroscopy (FT-IR) and Raman spectroscopy are excellent tools to identify and to study lipids. They allow easy identification of the components of mixtures, based on the assigned bands of standards (Table 6.8.1). In addition to the purely qualitative nature of the research, quantitative analysis can also be performed, and in the case of more advanced research, it is possible to use the relationship of the position of some well-known bands on the conformation of the lipid. This in turn pinpoints the exact state and environment in which the lipid is present – these properties are very useful when examining the behaviour of lipids in biological membranes.

In addition, both techniques allow determining important parameters concerning the quality of the lipid mixtures:

1. Presence of harmful *trans* isomers of C=C
   In Figure 6.8.2, a second derivative FT-IR spectra containing *trans* (blue) and *cis* (red) fats are shown. The characteristic band of *trans* fats is at 966 cm^{-1}, while for *cis* fats it is present around 961 cm^{-1} [2].
2. Degree of unsaturation of lipids
   The degree of unsaturation of lipids can be determined using mainly a band at about 3010 cm^{-1}, which originates from the vibration of a =C-H bonds connected with a C=C double bond. Figure 6.8.3 presents ATR FT-IR spectra of series of

Table 6.8.1. Band assignment of the major bands present in IR spectra of lipids [1]

Wavenumber [cm^{-1}]	Vibration	Assignment
3010	ν (=C-H)	stretching
2956	ν_s(-C-H)	asymmetric stretching CH$_3$
2920	ν_{as}(-C-H)	asymmetric stretching CH$_2$
2870	ν_{as}(-C-H)	symmetric stretching CH$_3$
2850	ν_s(-C-H)	symmetric stretching CH$_2$
1730	ν(C=O)	stretching C=O
1470	δ_{as}(-C-H)	asymmetric bending CH$_2$
1460	ν(-C-H)	scissoring CH$_3$
1400-1200	ρ(-C-H)	progression CH$_2$
1170	ν_{as}(-CO-O-C)	asymmetric stretching CO-O-C
1070	ν_s(-CO-O-C)	symmetric stretching CO-O-C
966	β(-C=C-)	deformation *trans* -C=C-
720	ω(C-H)	rocking CH$_2$

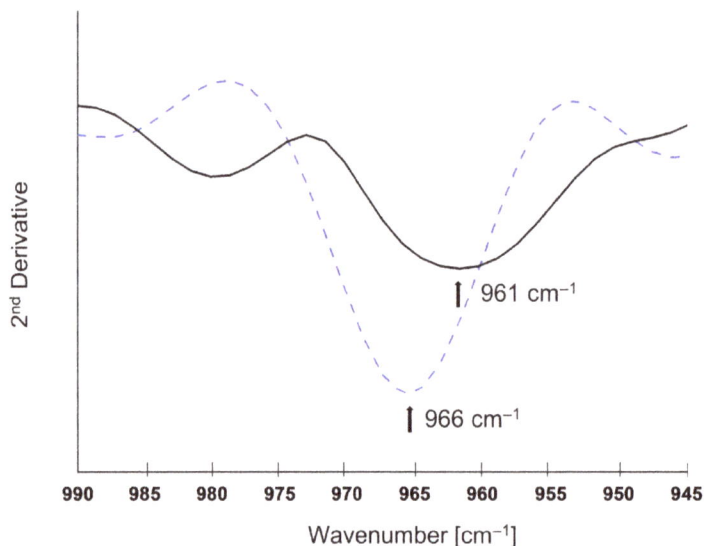

Fig. 6.8.2. The second derivative FT-IR spectra of mixtures containing *trans* (blue) and *cis* (red) fats. Detailed discussion in the reference [2]

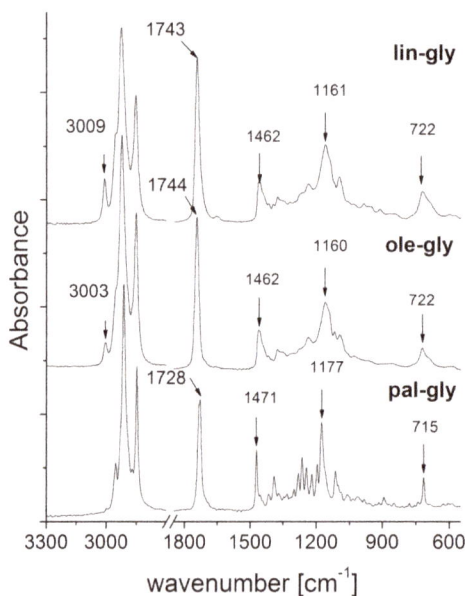

Fig. 6.8.3. ATR FT-IR spectra of a series of three triglycerides, differing in fatty acid residues, which are designated R1, R2 and R3 as in Figure 6.8.1 respectively, here: glyceryl tripalmitate (pal-gly), glyceryl trioleate (ole-gly) and glyceryl trilinoleate (lin-gly), from [3] – Reproduced by permission of The Royal Society of Chemistry

three triglycerides, differing in fatty acid residues – residues are sequentially designated R1, R2 and R3 as in Figure 6.8.1. The compound containing R1 has a spectral profile typical for unsaturated aliphatic chains, which do not contain vibrations of carbon-carbon double bonds. In the R2 case, having one double bond results in a band that can be observed at 3003 cm^{-1}, whereas in the case of R3 there are already two double bonds present and the intensity of the $\nu(=$C-H$)$ band at 3008 cm^{-1} is respectively greater.

3. The type of lipids present in a sample

Position of the carbonyl band to a fairly good degree indicates the lipid group, to which an analysed compound belongs to: 1690–1715 cm^{-1} fatty acids, 1730–1740 cm^{-1} cholesteryl esters and 1735–1750 cm^{-1} triglycerides. Unfortunately, the position of this band is also dependent on the conformation and aggregation state of the compound and thus also of polymorphic structure. For example, the compound containing R1, apart from not having carbon-carbon double bonds differs also from R2 and R3 by having a shorter aliphatic chain, missing two CH$_2$ groups. However, it is not this structural feature that is the cause of the carbonyl band shift to out of typical triglycerides range, *i.e.* 1728 cm^{-1} – this stems from the different state of matter, which for this compound is solid, while for R2 and R3 is liquid. This is additionally confirmed by the presence of so called CH$_2$ progression, which is seen as a series of spikes in the 1200–1300 range of glycerol tripalmitate.

Using the ratio of CH$_2$/CH$_3$ bands in the 2800-3000 cm^{-1} range it is also possible to conclude about the chain length of the fatty acid residues, and generally about the aliphatic chains branching. The high intensities of the bands derived from the vibration of the CH$_2$ groups (2920 and 2850 cm^{-1} bands – Table 6.8.1) indicate the presence of long chains, whereas signal coming from CH$_3$ groups vibrations (2956 and 2870 cm^{-1} bands) give information about the branching, since each branch needs to end with a CH$_3$ group and therefore their relative contribution in the spectrum increases.

AIM OF THE EXPERIMENT

1. An analysis of the naturally occurring lipids using infrared spectroscopy.
2. Differentiation of lipid subgroups based on their FT-IR spectra.
3. Interpretation of FT-IR spectra of samples of a biological origin.

SCIENTIFIC BACKGROUND

1. Fundamentals of absorption FT-IR spectroscopy; Chapter 1.
2. Construction and operation of FT-IR spectrometer, including the ATR technique; Chapter 3.1.
3. Basics of the protein structure.

EQUIPMENT, MATERIALS, CHEMICALS

1. Standards of substances such as cholesteryl palmitate, glyceryl tripalmitate, glyceryl tristearate, palmitic acid. Natural lipid-containing samples, e.g., oil, olive oil, butter, lard, nuts.
2. ATR FT-IR spectrometer, e.g. Bruker Alpha.
3. Software for spectra visualization, e.g. OPUS software.

PROCEDURE

1. Measure ATR FT-IR spectra of standards of different groups of lipids along with samples of biological origin – optimize the number of scans for background and sample to obtain an adequate signal to noise ratio in the shortest possible time.
2. Identify the characteristic bands in the spectrum of lipids on the basis of Table 6.8.1.
3. Identify the type of lipids present in the sample based on the assignment of the bands (CH_2/CH_3 and $C=O$ bands).
4. Identify the presence of unsaturation of lipids.
5. Determine if harmful *trans* isomers are present in the sample.

REPORT

1. Read band positions from the obtained spectra.
2. Identify and discuss the types of lipids and the length of the aliphatic chains.
3. Identify and discuss the presence of *trans* fats.
4. Identify and discuss the presence of unsaturated bonds.
5. Discuss the utility of the method for the analysis of lipids in food.

References

1. Stuart B.H., *Infrared Spectroscopy: Fundamentals and Applications*, John Wiley & Sons, Chichester, UK, 2004.
2. Mossoba M.M., Seiler A., Steinhart H., Kramer J.K.G., Rodrigues-Saona L., Griffith P., et al., *Regulatory Infrared Spectroscopic Method for the Rapid Determination of Total Isolated Trans Fat: A Collaborative Study*, J. Am. Oil Chem. Soc., **88**, 1 (2011).
3. Wróbel T.P., Mateuszuk L., Chlopicki S., Malek K., Barańska M., *Imaging of lipids in atherosclerotic lesion in aorta from ApoE/LDLR-/- mice by FT-IR spectroscopy and Hierarchical Cluster Analysis*, Analyst, **136**, 24 (2011).

6.9. Structural analysis of carbohydrates by means of FT-IR spectroscopy

Kamilla Malek, Ewelina Wiercigroch

6.9.1. Molecular structure of carbohydrates

Carbohydrates known also as sugars or saccharides are aldehydes (aldoses) and ketones (ketoses) containing several hydroxyl groups. They constitute majority of organic matter in the Earth and exhibit energetic function as well as participate in metabolic processes [1]. Moreover, ribose and deoxyribose are moieties in nucleic acids, RNA and DNA, respectively. Carbohydrates are composed of monosaccharides containing five (pentoses) or six (hexoses) carbon atoms. Formulas of

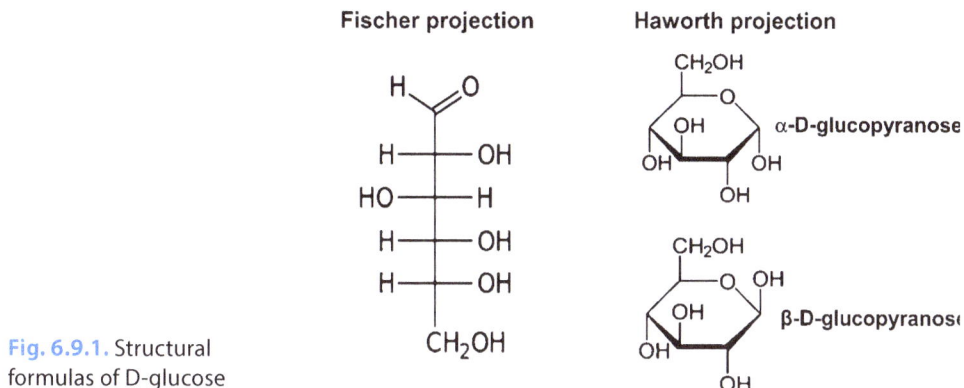

Fig. 6.9.1. Structural
formulas of D-glucose

MONOSACCHARIDES

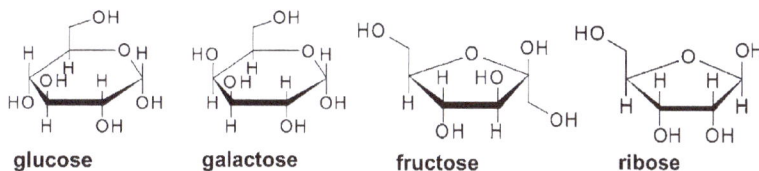

DISACCHARIDES

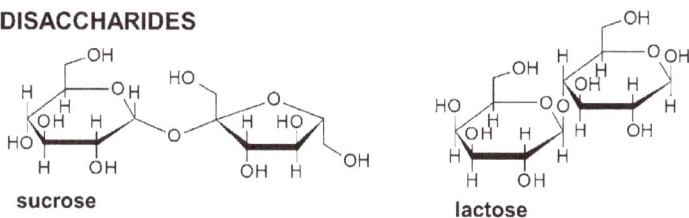

POLYSACCHARIDES

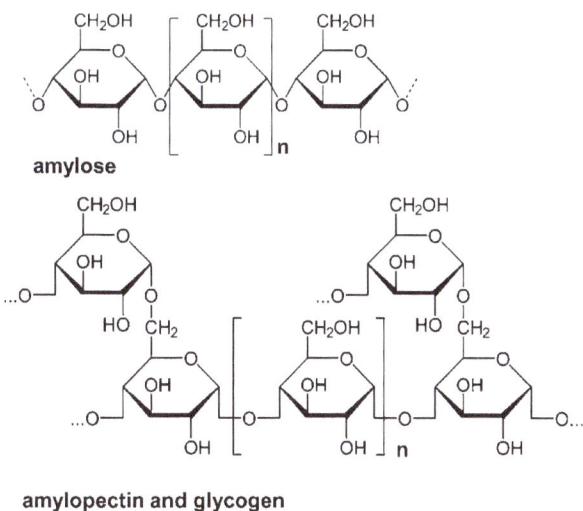

Fig. 6.9.2. Structural formulas of the selected carbohydrates

sugars are often presented in the Fisher projection. In solutions, the open-chain form (either "D-" or "L-") exists in equilibrium with several cyclic isomers, each containing a ring of carbons closed by one oxygen atom, as shown in Figure 6.9.1 for D-glucose. Carbohydrates also possess chiral centers, thus they are noted with symbols D, L, α i β to indicate spatial arrangement of the functional groups around the asymmetric carbon atom. Details are explained in Ref. [1].

Carbohydrates are divided into monosaccharides, oligosaccharides (composed of at least two monosaccharide units) and polysaccharides (polymeric carbohydrate molecules). Figure 6.9.2 illustrates molecular structures of the most common carbohydrates in nature, which are an object of this practical.

The most common monosaccharides in natural products are glucose, fructose and galactose belonging to hexoses as well as ribose being a pentose. In turn, oligosaccharides are formed from monosaccharides units joined together by the O-glycosidic bond. Starch consists of glucose and fructose residues *via* the α-1,2-O-glycosidic bond and therefore it shows a linear structure. In the case of lactose, galactose and glucose are bonded by the β-1,4-O-glycosidic bonding. Large polymeric oligosaccharides are formed from bonding of many monosaccharide residues. They usually play an important role in the energy storage in plants and animals. Plant-origin polysaccharide is starch consisting of two types of structures, namely amylopectin and amylose, in which glucose is a basic unit (Fig. 6.9.2). The non-branched amylose is formed by the α-1,4-glycosidic bond whereas in the branched amylopectin ratio between α-1,6-glycosidic and α-1,4-glycosidic bonds is approx. 1:30. Glycogen is a large-sized branched polysaccharide built of glucose. Similarly to amylopectin,

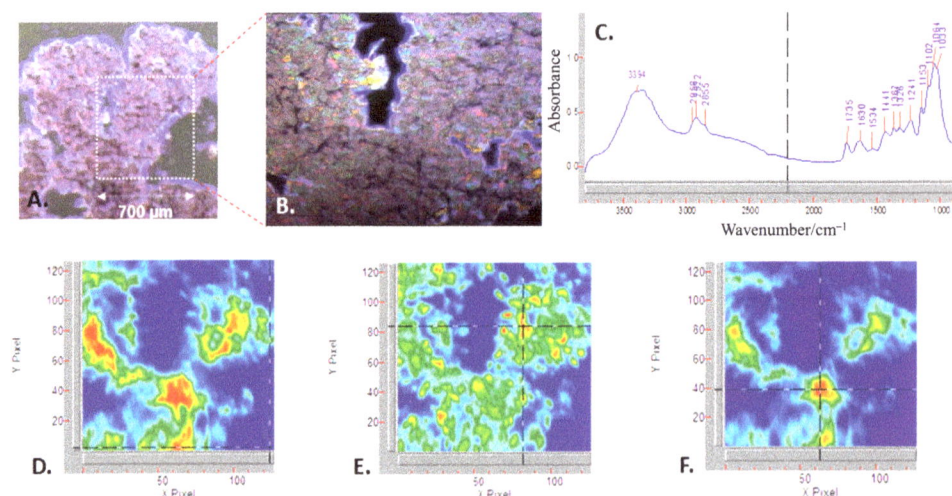

Fig. 6.9.3. Microphotography of a cross-section of a carrot sample: A. 4× magnification with an area imaged by FT-IR spectroscopy; B. 15× magnification; C. an exemplary FT-IR spectrum of carrot tissue; D. IR image for the distribution of sugars for the spectral region of 900-1200 cm^{-1}; E. IR image for the distribution of proteins for the amide I band; F. IR image for the distribution of the phosphate group for a band at 1240 cm^{-1}

molecules of glucose are joined through the α-1,4- and α-1,6-glycosidic bonds and the latter appears in each tenth bond.

6.9.2. Characteristic IR bands of carbohydrates

Functional groups of sugars are limited to CH, OH i COH groups and C-C bonds that exhibit high intensity bands in FT-IR spectra. From biological point of view sugars also participates in the formation of several biocomplexes such as glyco-proteins, proteoglycans, pectins, hemicelluloses, lignins, phosphosugars. Thus, IR spectra also show bands specific for sugars as well as for amide bands of pro-teins, the phenyl, phosphodiester, sulfate groups (see Chapter 6.9.3). An example of FT-IR analysis of chemical composition of a natural product is illustra-ted in Figure 6.9.3. FT-IR imaging of a cross section of carrot tissue (Amster-dam species) shows the distribution of carbohydrates along with proteins and the phosphate groups. Therefore, at regions of a high content of biocomponent one cannot exclude complexes between those macromolecules (Fig. 6.9.3 D-F, red color).

IR bands specific for mono-, oligo- i polycarbohydrates are mainly present in the spectral regions of 1450-800, 2800-3000 and 3100-3600 cm^{-1}. However, the high-wavenumber spectral range is also specific for proteins and lipids. Table 6.9.1 summarizes marker IR bands of sugars along with bands assignment. Individ-ual monosaccharides can be identified by IR bands below 930 cm^{-1} whereas di- and polysaccharadies are specified in IR spectra by a complex set of bands in the region of 1200 – 900 cm^{-1}.

Table 6.9.1. Positions of IR bands (in cm^{-1}) for sugars along with their assignment [2,3]

Spectral region	Assignment of bands
3650 – 3580	stretching of free OH groups; ν(OH)
3550-3200	stretching of H-bonded OH groups; ν(OH)
3000-2800	symmetric and asymmetric stretching of CH$_2$ and CH$_3$ groups; $\nu_{as}/_s$(CH$_3$/CH$_2$)
1750-1735	stretching of ester C=O groups in hemicelullose, pectins; ν(C=O)
1720-1705	stretching of C=O groups in acids; ν(C=O)
1630-1605	stretching of COO$^-$ groups in pectins; ν_{as}(COO$^-$)
1470-1430	scissoring of CH$_2$/CH$_3$ groups; δ(CH$_2$/CH$_3$)
1400-1330	bending of CH groups in ring; δ(CH)
1270-1200	bending of CH$_2$OH groups; β(CH$_2$OH)
1200-900	stretching of C-O and C-C; ν(CO/CC)
950-750	bending of COH, CCH and OCH; β(COH/CCH/OCH)

AIM OF THE EXPERIMENT

1. An analysis of molecular structure of natural sugars in FT-IR spectra.
2. Correlation of molecular structure of mono-, oligo- and polysaccharides with their marker bands in IR spectra.
3. Interpretation of FT-IR spectra of samples of a biological origin.

SCIENTIFIC BACKGROUND

1. Fundamentals of IR absorption spectroscopy; Chapter 1.
2. Construction and principle of operation FT-IR spectrometer including the ATR technique; Chapter 3.1.
3. Chemical structure and biological function of carbohydrates.

EQUIPMENT, MATERIALS, CHEMICALS

1. Reference substances: glucose, fructose (or other monosaccharides), saccharose, glukogen, starch and biological samples with a high-content of sugars.
2. FT-IR spectrometer with a ATR crystal (e.g. diamond).
3. Software for spectral analysis, e.g. Origin, Opus.

PROCEDURE

1. Collect ATR FT-IR spectra of reference substances, optimizing measurement settings (number of scans for background and sample) to record good signal to noise ratio in the shortest possible time and in the entire spectral range from 375 to 4 000 cm^{-1}.
2. Determine marker bands of mono-, di- and polysaccharadies in IR spectra of reference substances. In the case of overlapping bands, calculate second derivative spectra.
3. Measure ATR FT-IR spectra of natural products with a high content of carbohydrates.
4. Determine their marker bands based on IR spectra of reference substances, see Table 6.9.1.

REPORT

1. Indicate marker bands of mono-, di- and polysaccharides in their FT-IR spectra for reference substances and assign them to vibrations (Table 6.9.1). Discuss briefly similarities and differences between e.g. glucose and fructose, and then between both monosugars and sucrose.
2. Determine marker IR bands of carbohydrates in spectra of natural products, if possible identify sugar present in a sample. Discuss the presence of identified sugars with respect to the expected composition.
3. Discuss the utility of ATR FT-IR technique for the analysis of sugars in natural products.

References

1. Berg J.M, Tymoczko J.L, Stryer L., Biochemia, Wydania 5, PWN, 2002.
2. Li-Chan E.C.Y., Griffiths P.R., Chalmers J.M., *Applications of vibrational spectroscopy in food science*, Vol. 1, Wiley, 2010.
3. Schulz H., Barańska M., *Identification and quantification of valuable plant substances by IR and Raman spectroscopy*, Vib. Spectrosc., **43**, 13 (2007).

6.10. An analysis of ATR FT-IR spectra of animal tissues

Emilia Staniszewska-Ślęzak, Kamilla Malek

The application of Fourier-Transform Infrared spectroscopy in the analysis of biological samples like tissues dates back over half century. Nowadays this technique is commonly used, especially by using transmission and attenuated total reflection (ATR) modes in imaging of cross sections of various tissues, to find markers of a disease entity. However, there is also possibility to record single FT-IR spectra of tissue [1]. Infrared spectrum provides comprehensive information on biochemical components such as proteins, lipids, nucleic acids and carbohydrates, especially in the region of *"fingerprint"*. This enables the identification of spectral markers for each tissue as well as their differentiation in a pathological state.

6.10.1. Biochemical features of selected animal tissues

All tissues differ in their morphology, and consequently in their molecular composition. Bellow, selected animal tissues such as brain, lung, heart, liver, kidney and intestines are characterized in means of their general biochemical composition.

Brain consists of two types of tissues, grey and white matter. White matter contains a higher lipid (15%) and lower water (70%) content than the grey matter, whereas the latter is characterized by a lower lipid (5%) and higher water (83%) content. The concentration of proteins is similar for both of the matters, about 7.5–8.5 % [1,2].

A histological evaluation of the lung tissue identifies the presence of bronchial mucus, epithelium, fibrocollagenous stroma, smooth muscles, glandular tissue and cartilage. Since most of the lung tissues mainly exhibit protein composition, the total content of lipids does not exceed 2.9% [1,2].

The cardiac extracellular matrix contains fibrillar collagens and the primary structural proteins of the myocardium are actin and myosin [3]. In turn, karat parenchymal (hepatocytes) and non-parenchymal cells are the major compartments of liver. The total content of proteins, lipids, and nucleic acids is 12–15%, 5%, and 2%, respectively [1,4]. Microscopic sections of the intestinal tissue show the presence of serosa, muscularis externa, submucosa, muscularis mucosea, lamina propria and intestinal epithelium, which are built up protein filaments while kidneys participate in the whole body homeostasis through nephrons composed of coils and glomerulonephritis [1].

6.10.2. FT-IR spectra of animal tissues

As discussed above, each type of tissues exhibits different content of biocomponents such as proteins, lipids, nucleic acids and carbohydrates. This fact is reflected in FT-IR spectrum of an animal tissue. A typical spectrum of a biological sample is presented in Figure 6.10.1. The same set of IR bands is found in spectra of each tissue or a cell.

Second derivative IR spectrum depicted in Figure 6.10.2 emphasizes small differences in spectral band shapes and positions. It illustrates dominant absorption features of proteins, lipids, nucleic acids and carbohydrates. Second derivative spectra provide more detail description of tissue composition, for example they reveal secondary structure of proteins. Table 6.10.1 summarizes an assignment of IR bands of cells and tissues to particular vibrations of biomacromolecules.

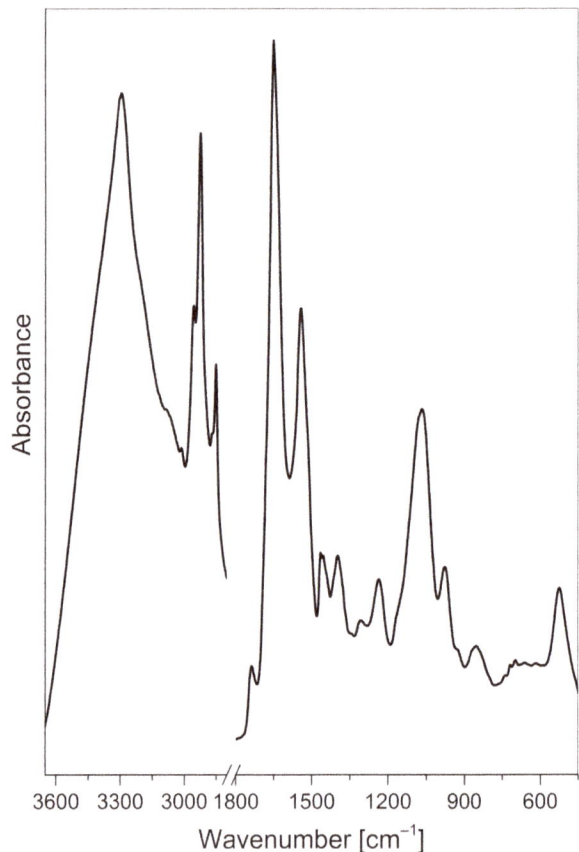

Fig 6.10.1. A typical FT-IR spectrum of a biological sample

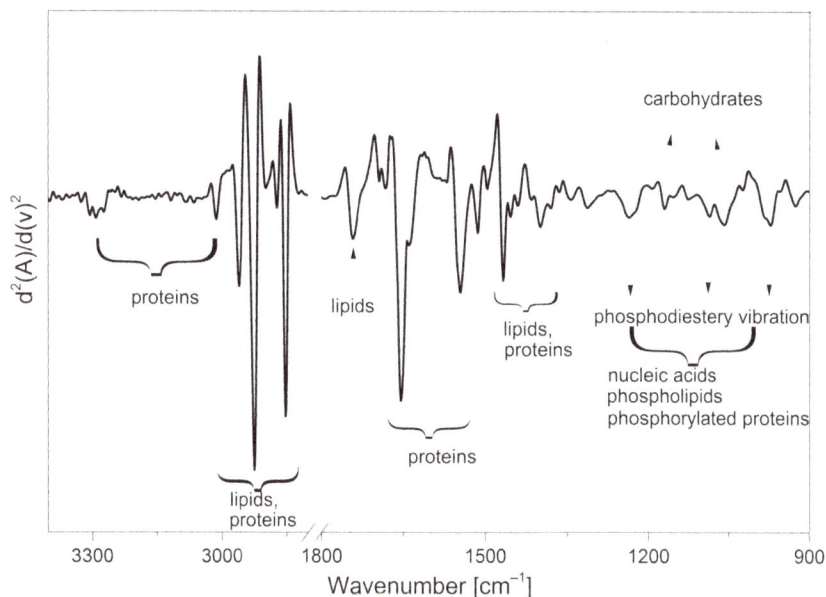

Fig. 6.10.2. Second derivative IR spectrum of a biological sample of an animal origin with labelled spectral regions specific for macromolecules

Table 6.10.1. An assignment of IR bands of animal tissues [4]

Vibrational mode	Wavenumber (cm^{-1})	Assignment
NH stretch (amide A)	ca. 3290	proteins
NH stretch (amide B)	3060	proteins
=CH olefinic stretch	3010-3030	unsaturated lipids
CH$_3$ asymmetric stretch	2954-2962	lipids and protein
CH$_2$ asymmetric stretch	2915-2924	saturated lipids and side chain of proteins
CH$_3$ symmetric stretch	2870	lipids and protein
CH$_2$ symmetric stretch	2847-2855	saturated lipids and side chain of proteins
SH stretch	2550	cysteine thiol group
C=O stretch (ester)	1740-1730	lipids, phospholipids
C=O stretch	1705-1690	bases of nucleic acids
C=O stretch + NH bend (amide I)	1695-1670	proteins – anti-parallel α-sheet and β-turn structures
	1670-1660	proteins – 3$_{10}$-helix structure
	1660-1648	proteins – α-helix structure
	1637-1623	proteins – β-sheet structure
ring CC stretch of Tyr and Phe residues, and nucleotides	1620	proteins, nucleic acids
NH bend + C-N stretch (amide II)	1580-1480	proteins

Table 6.10.1. (continued)

Vibrational mode	Wavenumber (cm⁻¹)	Assignment
ring CC stretch of Tyr residues	1520	tyrosine proteins
CH₂ scissoring mode	1473-1462	lipids proteins,
CH₂, CH₃ deformation modes	1455	proteins
COO⁻ symmetric stretch	1400	fatty acids, amino acids
CH₃ deformation of aliphatic side groups of amino acid residues	1379	proteins
CH₂ wagging	1343	phospholipids, fatty acids, triglycerides, amino acid side chains
amide III	1340-1240	proteins
PO₂⁻ asymmetric stretch	1238	DNA
CO-O-C asymmetric stretch	1171	cholesteryl esters
C-OH stretch	1171	serotine, tyrosine, threonine
C-O stretch	1151	glycogen, mucin
C-O stretch of ribose ring	1120	RNA
PO₂⁻ symmetric stretch C-C stretch	1083-1086 1078	nucleic acids, phospholipids glycogen
-CO-O-C stretches	1063	cholesterol esters, phospholipids
COH deformation	1050-1056	mucin, carbohydrates
	1030 1022-1028	nucleic acids glycogen, carbohydrates
Dianionic phosphate monoester	970	phosphorylated proteins
C-N-C stretch of ribose-phosphate skeletal vibrations	965	nucleic acids
	933	Z type DNA

A brief summary of IR bands for proteins, lipids, carbohydrates and nucleic acids is presented below.

Proteins

Proteins are the dominating constituents in all cells and tissues, and their vibrational fingerprint is easily observed in infrared spectroscopy (see Chapter 6.7). Protein IR spectrum has two primary features, called amide I and amide II bands, which are widely used in studies on protein secondary structures since both are sensitive to structural conformation. These vibrations assigned predominantly to the C=O stretching and NH bending modes are observed in the 1670 – 1620 and 1560 – 1530 cm⁻¹ regions, respectively. Other bands characteristic for proteins only give maxima in the high-wavenumber region above 3020 cm⁻¹ and they are attributed to the NH stretching vibrations (amide A and B). However, amide A and B are not usually treated as spectral biomarkers. The amide I band is most extensively used to identify and quantify the secondary structure of proteins. Amide bands at ca. 1665 and 1540 cm^{-1} signify protein α-helical chains, while a maximum of in the 1640-1620 cm^{-1} region indicates the presence of a β-sheet structure. Additionally, the appearance of a shoulder band at ca.1695–1670 cm^{-1} exhibits the contribution of anti-parallel β-sheets and turns into protein composition. A number of less pronounced bands of proteins appears in the region of 1460–1240 cm^{-1}, however vibrations of other biomolecules (lipids or nucleic acids) can contribute to these bands, see Table 6.10.1 for detail. Proteins containing tyrosine residues (Tyr) exhibit their presence by a band at ca. 1520 cm^{-1} [4].

Lipids

The concentration of lipids in tissues is much less than proteins. The most discernible spectral features of lipids or phospholipids originate from the stretching modes of the CH$_2$ and =CH groups (2800-3050 cm^{-1}), the ester C=O bond (*ca.* 1736 cm^{-1}) as well as from the scissoring vibrations of the long-chain CH$_2$ groups (*ca.* 1468 cm^{-1}). It should be emphasized that biochemical processes, in which lipid metabolism is strongly involved, are mostly recognized by the examination of the high-frequency region. There are usually four bands attributed to the asymmetric and symmetric stretching vibrations of the methyl and methylene groups that are found at $\sim$ 2960 [$v_{as}(CH_3)$], 2923 [$v_{as}(CH_2)$], 2873 [$v_s(CH_3)$], and 2850 cm^{-1} [$v_s(CH_2)$]. In addition, a weak band at ca. 3010 – 3030 cm^{-1} is associated with the CH stretching vibration of unsaturated fatty acids [$v(=CH)$]. The stretching vibration of the ester C=O group at ca. 1740 cm^{-1} is assigned to triglycerides as well as cholesterol esters, whereas bands at 1713 cm^{-1} indicates the presence of free fatty acids [4]. A detail description of lipids is given in chapter 6.8.

Carbohydrates

In general, the spectral range of 1150-1000 cm^{-1} is typical for carbohydrates. A maximum at ca. 1085 and 1050 cm^{-1} with a shoulder at around 1030 cm^{-1} is assigned to mono- and polysaccharides as well as proteoglycans. Additionally, three specific bands at 1029, 1078 and 1152 cm^{-1} are typical for glycogen [4].

Nucleic acids

FT-IR spectra also show a few characteristic bands for nucleic acids. They originate from the C=O and ring CC stretching modes of the purine and pyrimidine bases (in the region of 1720–1600 cm^{-1}), the asymmetric and symmetric stretches of the phosphate group as well as stretches of the C-O bonds in the phosphodiester and sugar moieties (in the region of 1300–900 cm^{-1}). Bands specific for DNA are found at1240 and 1080 cm^{-1} and they correspond to the asymmetric and symmetric stretching vibrations of the phosphate group. Additionally a feature at 1120 cm^{-1} is exclusively assigned to RNA.

AIM OF THE EXPERIMENT
1. Interpretation of FT-IR spectra of animal tissues.
2. Correlation of tissue chemical composition and marker bands in IR spectrum;

SCIENTIFIC BACKGROUND

1. Fundamentals of IR absorption spectroscopy; Chapter 1.
2. Construction and principle of operation FT-IR spectrometer including the ATR technique; Chapter 3.1.
3. Basic information about chemical composition of animal tissues and IR markers of major biomacromolecules.

EQUIPMENT, MATERIALS, CHEMICALS

1. Tissues samples delivered by students for the practical, e.g. chicken liver, kidney, chicken breast.
2. FT-IR spectrometer with a ATR crystal (e.g. diamond).
3. Software for spectra analysis, e.g. OPUS, Origin.

PROCEDURE

1. Measure ATR FT-IR spectra of tissue samples in the whole spectral range from 400 to 4 000 cm^{-1}.
2. Determine marker bands of proteins, lipids, carbohydrates and nucleic acids. Calculate second derivative spectra and indicate marker bands for each macromolecule.
3. Based on second derivative spectrum determine bands specific for secondary structures of proteins (see Chapter 6.7).
4. Calculate integral intensity of the following bands: amide I, $\nu_{as}(CH_2)$, $\nu_{as}(PO_2^-)$, bands in the range of 1050–1000 cm^{-1}.

REPORT

1. Based on Table 6.10.1 indicate marker bands for proteins, lipids, nucleic acids and carbohydrates in IR spectra of tissues and assign them to vibrations.
2. Determine secondary structure of proteins in the studied tissues.
3. Based on integral intensity, calculate ratios of proteins/lipids, proteins/carbohydrates, proteins/nucleic acids. Comment these results in the context of the biochemical composition of an analysed tissue.
4. Discuss the utility of ATR FT-IR method for the analysis of animal tissues.

References

1. Staniszewska E., Malek K., Barańska M., *Rapid approach to analyze biochemical variation in rat organs by ATR FTIR spectroscopy*, Spectr. Chim. Acta A, **118**, 981 (2014).
2. Krafft C., Sergo V., *Biomedical applications of Raman and infrared spectroscopy to diagnose tissues*, Spectrosc. **20**, 195 (2006).

3. Movasaghi Z., Rehman S., Rehman I., *Fourier transform infrared (FTIR) spectroscopy to biological tissues,* Appl. Spectrosc. Rev., **43**, 134 (2008).
4. Malek K., Wood B.R., Bambery K., in: *Optical Spectroscopy and Computational Methods in Biology and Medicine* (ed. M. Baranska) Springer, 2014.

6.11. Diagnostics of disease development by FT-IR imaging of tissue

Kamila Kochan, Małgorzata Barańska

The possibility of obtaining information about the biochemical composition on the molecular level makes FT-IR spectroscopy an excellent tool for the study of the pathogenesis of diseases, providing information not available by other techniques. A huge diagnostic potential of FT-IR spectroscopy results not only from the above, but also from a number of advantages like simplicity of sample preparation and performance of the measurements as well as the non-destructiveness of this technique. In the case of the application of spectroscopic techniques (including FT-IR spectroscopy) to the study of animal tissues it is important to be aware of the complex composition of the research subjects. This means in practice that any signal (e.g. the band located at 1656 cm^{-1}) is attributed more likely to a group of compounds, rather than to a particular substance. This, however, still allows to draw conclusions *i.a.* about the global content of compounds and their mutual relationship (e.g. ratios: proteins/lipids, proteins/DNA, lipids/DNA) or their detailed structure (e.g. the secondary structure of proteins). A detail characteristics of the IR bands visible in spectra of biological samples is described in chapters 6.7 – 6.10.

In this type of studies animal models of diseases are most often used. Diseases can be induced by genetic modification (e.g. mice ApoE/LDL$^{-/-}$ – a model of atherosclerosis, db/db mice – a model of diabetes), environmental factors such as diet (e.g. C57BL/6J on a High Fat Diet – a model of Non-Alcoholic Fatty Liver Disease) or a combination of these (mixed models, e.g. ApoE/LDL$^{-/-}$ mice on the Low Carbohydrate High Protein diet – a model of advanced atherosclerotic lesions).

Biological samples for FT-IR measurements usually do not require any special preparation. Fresh tissue should be measured as soon as possible after their preparation, due to the possibility of changing their composition, caused by the action of factors such as light or temperature. An alternative approach, allowing for longer storage of samples and protection against the destructive influence of the environment, is to preserve samples e.g. with 4% buffered formalin solution.

AIM OF THE EXPERIMENT

The aim of this exercise is to demonstrate the diagnostic potential of FT-IR spectroscopy for the detection and determination of the degree of the disease development. Application of FT-IR imaging for diagnosis includes two key elements: (1) measurements as well as (2) an analysis of the obtained data. The second step is particularly important in the case of the early stages of disease, characterized by a low degree of changes occurring due to pathological alterations.

In order to use the FT-IR imaging to assess the development of the disease it is necessary to perform in the first place (1) the analysis of the biochemical composition of samples of biological material (tissue): healthy and pathologically changed, with lesions at various stages of development. By comparing their composition it is possible to (2) determine the trends of biochemical modifications caused by the disease. Knowledge of the biochemical composition of the tissue as well as about changes emerging due to the development of the disease allows then for (3) its detection and determination of the stage of its development in samples of unknown condition.

SCIENTIFIC BACKGROUND

1. Theoretical basics of IR absorption spectroscopy; Chapter 1.
2. The design and principle of operation of FTIR spectrometer with regard to digital imaging; Chapter 3.2.
3. The basics of tissue composition: ranges of occurrence of bands characteristic for the main biomolecules; Chapter 6.10.
4. Principles of chemometric analysis; Chapter 5.

EQUIPMENT, MATERIALS, CHEMICALS

1. In this exercise tissue samples will be used. Tissue sections of chosen thickness (e.g. 10 μm) are placed on a window made of a material transparent to infrared radiation (e.g. CaF_2) and subsequently fixed with 4% buffered formalin solution.
2. For measurements, a FT-IR spectrometer equipped with a IR Cassegrain microscope (15×) and a FPA detector (e.g. Agilent 670) will be used. Parameters such as spectral resolution and the number of scans will be chosen experimentally to ensure a high signal to noise (S/N) ratio of the obtained data in minimal acquisition time. These parameters will be kept constant for all of measurements, in order to allow comparison of results and to minimize their impact on the observed spectral differences.
3. Software for the analysis of FTIR imaging results (e.g. CytoSpec) as well as for the analysis of individual spectra (e.g. Opus, Origin).

PROCEDURE
I. FTIR imaging
   All measurements should be performed in transmission mode.
   1. Determine the experimental parameters such as spectral resolution (typically: 8 cm^{-1}) and a number of scans for background (e.g. 128) and sample (e.g. 64). These parameters remain fixed in further measurements.
   2. Collect background by measuring a fragment of a window without the sample.
   3. Collect two FT-IR images for two control samples (one for each).
   4. Collect two FT-IR images for each of the samples at different known stages of the disease development.
   5. Collect one FT-IR image of samples with unknown pathological state.

II. Data analysis
   1. Open the measured FT-IR images in software allowing the visualization of digital imaging results (e.g. CytoSpec) and:
      a. perform an assignment of characteristic bands of lipids, proteins, carbohydrates and DNA for chosen spectrum as shown in Table below;
      b. create images showing the distribution of the selected components based on the integration of the bands characteristic for proteins (e.g. amide I band), lipids, carbohydrates and DNA;
      c. calculate averaged spectrum from each image.
   2. In Opus software (or other) calculate the integral intensity of the selected bands (bandwidth will be chosen depending on the type of tissue/disease during class) for the average spectra.

REPORT
1. Compare the distribution of the major components (proteins, lipids, carbohydrates, DNA) and averaged spectra from the images collected from control tissues and discuss:
   a. are there any differences between them?
   b. what can be the reason for that?
   c. is the presence of differences between samples of the same type justified?
2. Compare the results for samples at the different (known) stage of the disease with the results for control samples by comparing the images showing the distribution of the main components (proteins, lipids, carbohydrates, DNA) – are there any specific changes emerging along with the development of the disease (e.g. aggregation or disappearance of the selected components)?
3. On the basis of the calculated integral intensity determine the relation between the main components of tissue:

Ratio	Control samples	Early stage of disease	Advanced stage
Lipids*/Proteins**			
Proteins/DNA***			
Lipids/DNA			

* total lipid content can be expressed as a sum of bands: $v_{as}(CH_2)$ and $v_s(CH_2)$ (~2982 cm^{-1} and ~2873 cm^{-1});
** total protein content an be expressed as a sum of amide I and II band (~1650 cm^{-1} and ~1555 cm^{-1});
*** total DNA content can be measured by the band $v_s(PO_2^-)$ (~1084 cm^{-1})

4. On the basis of the analysis carried out in 2 and 3 specify:
   a. the difference between the control (healthy) and known pathologically changed samples (e.g. in the distribution of individual components, their contents relative to each other, the secondary structure of the proteins, *etc.*)
   b. trend (or lack of it) of changes with the development of the disease (e.g. increase or decrease in the content of the group of compounds, changes in protein structure, *etc.*).
5. For samples with unknown stage of development of the disease carry out an analysis similar to that in 2 and 3.

6. On the basis of the analysis and the comparison of the spectra of the control samples and samples with defined degree of the disease, as well as the use of conclusions in the step 4 determine the status of samples.

Selected applications of Raman spectroscopy

7.1. The identification of proteins secondary structure in Raman spectra

Anna Rygula

An analysis of proteins by means of Raman spectroscopy (RS) is used primarily for two purposes: 1) to determine structure of proteins and 2) to provide information about proteins in an analysed biological material. The structure of proteins and their subdivisions together with their analysis in FT-IR spectroscopy is described in Section 6.7. In addition, heme proteins and their analysis with resonance Raman effect is presented in Section 7.9.

Both the primary and secondary structures of proteins can be analysed by using Raman spectroscopy. In the case of the proteins' primary structure, this technique can be applied to determine the presence of certain amino acids (e.g. aromatic or cysteine residues) but is not suitable to perform a sequence analysis of a protein chain. However RS can be used to determine the contribution of secondary structures such as helical (α), β-sheet and disordered as well as to track their changes in various physico-chemical conditions. Indirectly Raman techniques can be helpful in an analysis of a higher protein structures and local changes in certain amino acids [1-3].

An important advantage of Raman spectroscopy is its ability to analyse proteins in their native state without matrix isolation, which is required for other techniques. This is important because any change in the environment, e.g. pH or a solvent, can affect the secondary structure of proteins. Only understanding and analysis of the protein structure in its natural environment, e.g. in cells and tissues, gives reliable knowledge about protein function.

7.1.1. Characteristics of protein in Raman spectrum

Atoms forming the peptide bond are reflected in Raman spectrum by amide bands called A, B, I-VII (see Chapter 6.7). The presence of these bands distinguishes spectrum of an amino acids mixture from Raman spectrum of polypeptide [2]. Spectral ranges of amide bands and their description are presented in Table 7.1.

Table 7.1.1. Spectral features of amide bands including an assignment of the dominant vibrations [3]

Amide band	Spectral range [cm⁻¹]	Band assignment*
A	3250-3300	ν(N-H)
B	3030-3100	ν(N-H)
I	1600-1690	ν(C=O)
II	1480-1580	ν(C-N) δ(N-H)
III	1220-1300	ν(C-N), δ(N-H), ν(CH$_3$-C)
IV	625-700	δ(O=C-N)
V	640-800	γ(N-H)
VI	540-600	γ(C=O)
VII	~200	skeletal

*ν – stretching vibration, δ – bending vibration in plane, γ – bending vibration out of plane

Determination of secondary structures by means of Raman bands positions is not as accurate as by crystallography or circular dichroism (CD) (Chapter 3.3), but allows to determine a general trend, because the form and strength of hydrogen bonding C=O⋯H-N in amide bond shifts maxima of amide modes. To determine the secondary structure by Raman spectroscopy we primarily observe a shift of amide I and III bands in contrast to FT-IR and UV RRS techniques (mainly amide I and II bands) [2]. Positions of I and III amide bands for various protein structures are shown in Table 7.1.2 [2,3].

Table 7.1.2. Positions of I and III amide bands (in cm⁻¹) in Raman spectra for various protein secondary structures [2,3]

protein structure	amide I	amide III
α-helix	1655-1662	1264-1272
β-sheet	1672-1674	1227-1242
disordered	1665-1668	1239-1254

Because proteins are huge polypeptides consisting of hundreds of amino acids, their Raman spectra are composed of many overlapping bands. Moreover, only a few proteins exhibit a clearly defined structure as α or β, with no other contribution. An exemplary spectrum of proteinase (Subtilisin Carlsberg) is presented in Figure 7.1.1 and it belongs to the class of α/β proteins. In addition Raman spectra also show the presence of bands specific for amino acid residues. They are discussed below.

7.1.2. Marker bands of amino acid residues

The amino acid residues of the polypeptide chain allow examining local changes in the protein structure due to environmental stress like denaturation, conformational

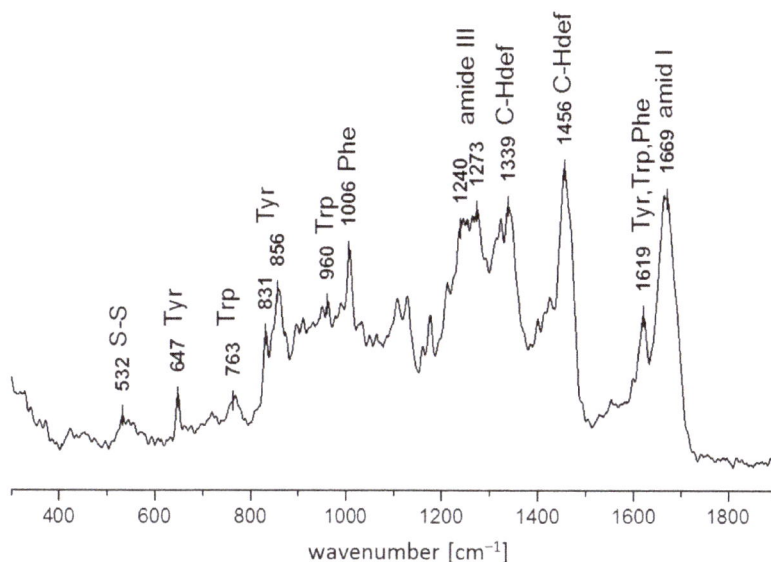

Fig. 7.1.1. Raman spectrum of proteinase (Subtilisin Carlsberg) with an assignment of bands to vibrations of protein constituents

changes, etc. The most useful spectral features for this purpose are bands attributed to residues of aromatic amino acids: tyrosine (Tyr), tryptophan (Trp), phenylalanine (Phe), histidine (His) and amino acids containing a sulphur atom like cysteine (Cys) and methionine (Met).

Tryptophan

The tryptophan bands are sensitive to microenvironment around this amino acid residue. This is determined by intensity ratio of bands at 1360 and 1340 cm^{-1} (so called a tryptophan doublet, I_{1360}/I_{1340}). The ratio expresses the hydrophobic/hydrophilic environment of the tryptophan residue, *i.e.* the higher value is, the more hydrophobic environment exists [2].

Tyrosine

Bands at 860 and 833 cm^{-1} originate from vibrations of the tyrosine residues and they are sensitive to the presence of hydrogen bonds. If the intensity ratio (I_{860}/I_{833}) is approx. 2.5, the OH group of Tyr is an acceptor in strong hydrogen bonding. In contrast, when the OH group plays a role of donor, this ratio is equal to approx. 0.3.

Histidine

The stretching vibration of the $C_4=C_5$ in the histidine ring appears in the range of 1574–1587 cm^{-1}. Wavenumber of this band is specific for the type binding of metal ions such as Cu(II) or Zn(II) to the His residue. This also allows determining structure of aggregates formed due to protein denaturation upon binding metal ions. This found application in studies on prion disease. Additionally, a band at 1408 cm^{-1} is indicative for His residues protonation, however, it is only observed in D$_2$O solution [2].

Phenylalanine

The presence of this amino acid is confirmed by a very sharp and intense band at 1005 cm^{-1}.

Vibrations of disulphide bridge

The disulphide bridge (R-S-S-R) plays a key role in the formation of protein tertiary structure. A Raman band of the S-S bridge is sensitive to its conformation, and thus it is helpful in the identification of protein structure. A position of the S-S stretching vibration, v(S-S) depends on the dihedral angle C$_\alpha$-C$_\beta$-S-S-C$_\beta$-C$_\alpha$. So, this band occurs at 508–512, 523–528, and 540–545 cm^{-1} for GGG, TGG, TGT conformations, respectively (T-*trans*, G-*gauche*). It should be mentioned here that Raman spectroscopy is one of a few techniques allowing for the determination of structures of disulphide bridges [2].

Cysteine and methionine

Cysteine and methionine include the -CS- moiety, which the stretching vibration band, v(C-S), appears in Raman spectrum at approx. 630–750 cm^{-1}. This vibration is sensitive to conformation of the side chain of these amino acid residues and their spatial arrangement with respect to the main chain of protein [1,2].

7.1.3. Other techniques of Raman spectroscopy for the analysis of proteins

The other popular method of spectroscopic analysis of proteins is resonance Raman spectroscopy in UV range (Ultraviolet Resonance Raman Spectrometry, UVRRS). This method relies on resonance effect, derived from UV absorption of the peptide backbone and aromatic amino acid residues in the UV region [1]. Therefore, Raman spectrum provides information about vibrations of the polypeptide chain and its changes due to environmental stress or coordination of metal ions.

Optically active Raman spectroscopy (called Raman Optical Activity, ROA, *see* Chapter 4.3) is used for an analysis of chiral molecules which interact with circularly polarized light. This technique is developed towards the identification of three-dimensional structure of proteins; however, the practical and routine use of this technique is still in progress. Importantly, it provides information about protein structure in solution [4].

The other extremely sensitive technique is surface-enhanced Raman spectroscopy (SERS, see Section 4.2). Its fundamentals rely on interaction of an analyte with a metallic surface, e.g. silver or gold colloid. The Raman signal recorded with this technique is significantly enhancement by a factor of 10^6-10^8. SERS is used both to determine very low concentrations of proteins (around 10^{-10} M) and to define their conformations [4].

AIM OF THE EXPERIMENT
1. An analysis of naturally occurring proteins by using Raman spectroscopy;
2. Correlation between protein structure and marker bands in Raman spectrum;
3. Interpretation of Raman spectra of samples with a biological origin.

SCIENTIFIC BACKGROUND
1. Fundamentals of Raman spectroscopy; Chapter 2.
2. Construction and operation of a Raman spectrometer; Chapter 2.
3. Secondary structures of proteins, Chapter 6.7.

EQUIPMENT, MATERIALS, CHEMICALS
1. Standards of proteins with α-helical (e.g. albumin, collagen), β-sheet (e.g. trypsin, chymotrypsinogen) or mixed structures (e.g. papain, α-lactalbumine), a sample of a biological origin containing a defined type of protein.
2. Raman spectrometer.
3. Software for spectral analysis, e.g. OPUS, Origin.

PROCEDURE
1. Measure Raman spectra of reference substances.
2. Determine and indicate protein marker bands in collected spectra.
3. Measure Raman spectrum of a biological sample.
4. Indicate positions of marker bands.

REPORT
1. From the reference spectra indicate marker bands for proteins and assign them to amide and side chains vibrations. Determine secondary structure of proteins (α, β and α/β). Discuss differences and similarities in spectra of proteins with different secondary conformations (α and β).
2. Assign amide bands in Raman spectrum of a biological sample and determine secondary structure using the reference spectra from 1) and 2). Discuss the presence of an identified structure of a protein with expected conformation reported by the literature.
3. Discuss the application of Raman spectroscopy to study the structure of proteins and their identification in samples of a biological origin.

References

1. J. Twardowski, *Biospektroskopia*, vol. 4, 1990, PWN.
2. Kitagawa T., Hirota S., in: *Handbook of vibrational spectroscopy*, (eds. J.M. Chalmers, P.R. Griffiths), John Wiley, 2002,3426.
3. Ryguła A., Majzner K.,. Marzec K. M, Kaczor A., Pilarczyk M., Barańska M., *Raman spectroscopy of proteins: a review*, J. Raman Spectrosc., **44**, 1061 (2013).
4. Tuma R., *Raman spectroscopy of proteins: from peptides to large assemblies*, J. Raman Spectrosc., **36**, 307 (2005).

7.2. Raman analysis of fatty acids

Aleksandra Jaworska, Małgorzata Barańska

7.2.1. Occurrence and characteristics of fatty acids

Fatty acids are monocarboxylic acids with the general formula R-COOH. The hydrocarbon chains are simple, with an even number of carbon atoms, excluding the -COOH group. Fatty acids are often described by notation of $n:m$, where n is the number of carbon atoms in the molecule (including the C atom in the carboxyl group) and m is the number of double C=C bonds. In nature, these acids are present in the form of esters of glycerol commonly called fats. Fat consumed in adequate amounts is necessary for the proper functioning of the human body, but the type of fat is of great importance. Soft and liquid fats (oils and margarines produced from oils) are rich in poly- and monounsaturated fatty acids (SFA), and these acids are beneficial for health. In contrast, solid fats (butter, lard and hard margarines) contain a high amount of saturated fatty acids and *trans* isomers known from their adverse action on human health.

Detailed description of the common fatty acids is given below. Figure 7.2.1 and Table 7.2.1 illustrate structure and content in oils common in the use:

1. Polyunsaturated fatty acids:
    - actively reduce cholesterol level in blood;
    - occur mainly in liquid and soft fats such as vegetable oils (soybean and corn oils), soft margarine oils and various nuts;
    - among the polyunsaturated fatty acids we can distinguish essential fatty acids (EFAs), which are involved in the construction of cell membranes and the production of hormones. They can be divided into two groups: the family of linoleic acid (omega-6) and the family of α-linolenic acid-LNA (omega-3). EFAs from the family omega-3 exhibit high biological activity and the importance for the organism;
    - the human body cannot produce EFAs acids and therefore they must be supplied with food. The quantitative ratio of linoleic acid to linolenic acid is very important in diet, since the latter is not still consumed in enough amounts (Fig. 7.2.1, Table 7.2.1).

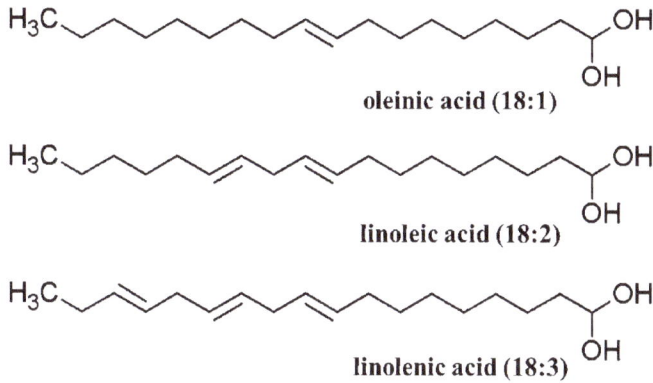

Fig. 7.2.1. The molecular structure of the selected fatty acids

Table 7.2.1. The approximate percentage of unsaturated fatty acid in selected oils [1,2]

Oil	oleinic acid (18:1) omega-9	linoleic acid (18:2) omega-6	linolenic acid (18:3) omega-3
corn	34	48	1
linseed	19	15	55
olive	74	9	
colza	15-60*	14-24*	8
sesame	40	42	
sunflower	14-39	48-74	3
grape	15	78	
peanut	61	19	

*percentage of fatty acids varies depending on oilseedtype colza.

2. Monounsaturated fatty acids:
    - are beneficial to health;
    - are mainly present in liquid fats, e.g. olive, sunflower and rapeseed oils as wells as in fats of soft consistency (soft margarines prepared from the listed oils).
3. Saturated fatty acids:
    - increase the level of cholesterol in blood, and therefore they cause a high risk of atherosclerosis and cardiovascular disease;
    - are mainly found in animal products such as meat, butter, milk fat, cheese, processed foods, and partially hydrogenated bakery fats.
4. *Trans*-fatty acids:
    - increase the level of cholesterol in blood;
    - occur in processed food and partially hydrogenated bakery fats (e.g. hard margarines, cookies, crackers, chips and crisps). A small amount of trans fats can be also found in milk, butter and beef.

7.2.2. An application of Raman spectroscopy in an analysis of fatty acids

Raman spectroscopy is an excellent tool for an analysis of fats. It allows specifying two parameters, important for their quality:

The presence of trans/cis isomers.

Fig. 7.7.2 depicts Raman spectrum of fat. Positions of bands typical for fatty acids along with the assignment of vibrations to the functional groups of fatty acids are listed in Table 7.2.2 [3].

The inset in Fig. 7.2.2 shows the presence of a band at 1660 cm^{-1} attributed to the stretching vibration of the C=C bond in the cis configuration with a shoulder

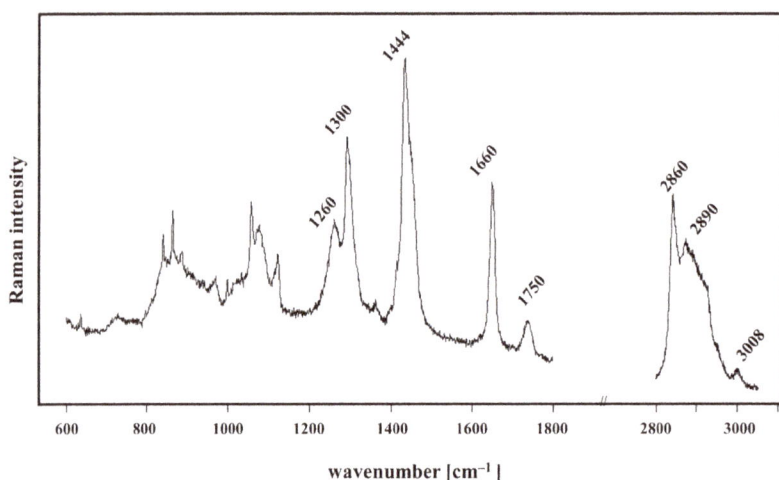

Fig. 7.2.2. Raman spectrum of fat

Table 7.2.2. The assignment of major bands present in Raman spectrum of fatty acid [3]

Wavenumber [cm^{-1}]	Functional group	Assignment	Description of vibration
3008	RCH=CHR	v_{as}(=C-H)	asymmetric stretching
2970	-CH$_3$	v_{as}(-C-H)	asymmetric stretching
2940	-CH$_2$	v_{as}(-C-H)	asymmetric stretching
2890	-CH$_3$	v_s(-C-H)	symmetric stretching
2860	-CH$_2$	v_s(-C-H)	symmetric stretching
1750	RC=OOR	v(C=O)	stretching
1670	*trans* RCH=CHR	v(C=C)	stretching
1660	*cis* RCH=CHR	v(C=C)	stretching
1444	-CH$_2$	δ(C-H)	scissoring
1300	-CH$_2$	τ_{ip}(C-H)	in-phase twisting
1266	*cis* RCH=CHR	δ_{ip}(=C-H)	symmetric wagging
1100-800	-(CH$_2$)$_n$	v(C-C)	stretching

at 1670 cm^{-1} originating from the trans isomer of fatty acid. This spectrum exhibits a typical Raman profile of fatty acid.

Degree of unsaturation of fatty acids.

The degree of unsaturation of fatty acids/fats is expressed by the iodine value (or iodine number), which is an amount of iodine, in grams, that is required to saturate C=C bonds in 100 grams of oil, fat, or wax:

$$\text{Iodine Number (iodine value)} = 100\text{g/fat mass} \times w \times M_{12},$$

where w – the number of double bonds in 1 mole of fat, M_{12} – molar mass of iodine molecule, fat mass – molar mass of fat present in the sample.

Drying oils used in the paint and varnish industry have relatively high iodine values (about 190). Semidrying oils, such as soybean oil, have intermediate iodine values (about 130) whereas Non-drying oils (olive oil, oils used for soap making and in food product) exhibit relatively low iodine values (about 80). In general, Raman spectroscopy allows calculation of the ratio between the number of the C=C to C-C bonds present in fatty acids [4,5], and therefore estimation of the iodine number. For this purpose, the ratio of the intensity of bands at 1660 cm^{-1} and 1444 cm^{-1} (I_{1660}/I_{1444}) can be used. Such an analysis of numerous fats showed a linear correlation between the I_{1660}/I_{1444} ratio and iodine number [1]. The level of unsaturation can be also determined from the ratio of the two other bands, at 1266 cm^{-1} and 1300 cm^{-1} (I_{1266}/I_{1300}) [2]. In addition, a band at 3008 cm^{-1} provides useful information on the presence of the unsaturated C=C groups in fatty acids [6].

Figure 7.2.3 shows FT-Raman spectra of three selected oils: olive oil (a), sunflower oil (b) and linseed oil (c). The comparison of the intensity ratio of bands I_{1660}/I_{1444} and I_{1266}/I_{1300} clearly exhibit an order of an increasing content of unsaturated fatty acids: olive oil < sunflower oil < linseed oil

These results well correlate with data collected in Table 7.2.1.

AIM OF THE EXPERIMENT

1. An analysis of naturally occurring fatty acids by using Raman spectroscopy;
2. Correlation of Raman profile and unsaturation of fatty acid and its isomers.

SCIENTIFIC BACKGROUND

1. Theory of normal Raman effect; Chapter 2.
2. Construction of the FT-Raman and dispersive Raman spectrometers; Chapter 2.
3. Structure and occurrence of fatty acids.

EQUIPMENT, MATERIALS, CHEMICALS

1. Natural samples containing fatty acids such as butter, oils, olive oil, margarines, nuts, avocado etc.
2. Fourier-transform Raman spectrometer.
3. Softwares to analyse spectra, e.g. Opus, Origin.

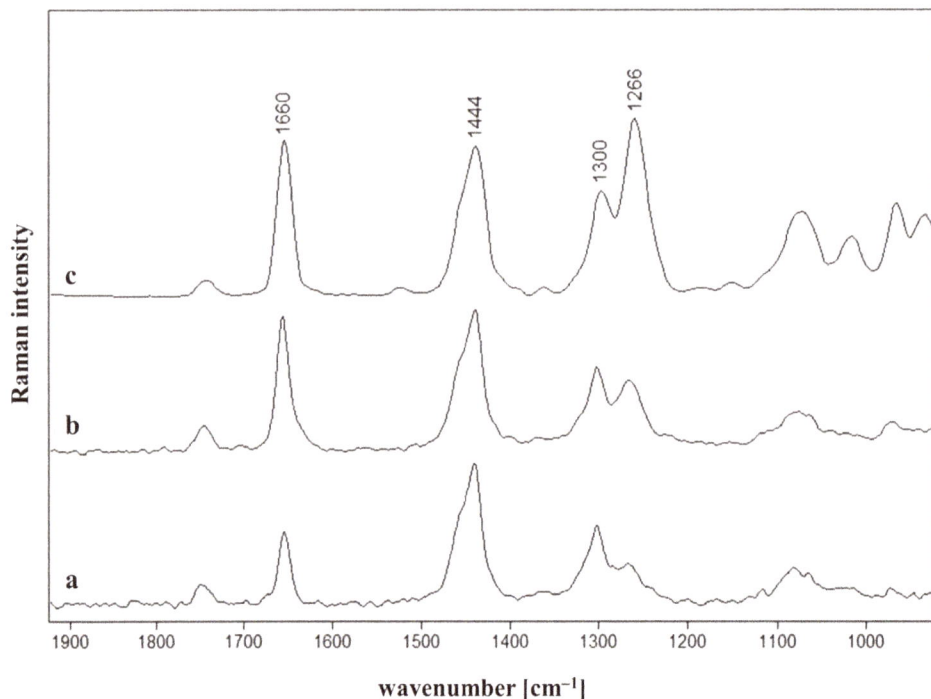

Fig. 7.2.3. FT-Raman spectra of olive oil (a), sunflower oil (b) and linseed oil (c)

PROCEDURE

1. Recording Raman spectra of samples brought by students, including optimizing measurement conditions (number of scans per sample, laser power) to obtain good signal to noise ratio in the shortest possible time in the spectral range from 0 to 4000 cm^{-1}.
2. Identification of Raman bands of fatty acids present in spectrum of a sample (Table 7.2.2).
3. Determination of degree of unsaturation of fatty acids by integration of selected Raman bands described above.
4. Identification of, if any, *trans* isomers in a sample.

REPORT

1. Read bands positions of fatty acids in recorded Raman spectra of a sample.
2. Integrate marker bands (I_{1660} and I_{1444} or I_{1266} and I_{1300}).
3. Calculate the ratio of I_{1660}/I_{1444} or I_{1266}/I_{1300}.
4. Order products with an increasing value of iodine number.
5. Identify the isomers, if the *trans* isomers are present in the sample?
6. Discuss the usefulness of Raman spectroscopy in an analysis of fatty acids in natural products/food.

References

1. M.A. Strehle et al., *Controlling the Quality of Different Vegetable and Fish Oils by Means of Raman Spectroscopy*, Proceedings XIXth International Conference on Raman Spectroscopy, P.M. Fredericks, R.L. Frost, L. Rintoul (eds.), CSIRO Publishing 2004, 411-412.
2. D. Lin-Vien et al., *The Handbook of Infrared and Raman Characteristic Frequencies of Organic Molecules*, Academic Press Inc., San Diego 1991.
3. B. Schrader, *Raman Spectroscopy in the Near Infrared – a most Capable Method of Vibrational Spectroscopy*, Fresenius J. Anal. Chem., **355**, 233 (1996).
4. R.C. Barthus et al., *Determination of the Total Unsaturation in Vegetable Oils by Fourier Transform Raman Spectroscopy and Multivariate Calibration*, Vib. Spectr., **26**, 99 (2001).
5. B. Muik et al., *Direct, Reagent-free Determination of Fatty Acid Content in Olive Oil and Olives by Fourier Transform Raman Spectroscopy*, Anal. Chim. Acta, **487**, 211 (2003).
6. V. Baeten et al., *Oil and Fat Classification by FT-Raman Spectroscopy*, J. Agric. Food Chem., **46**, 2638 (1998).

7.3. Raman spectroscopy as a method to analyze lipids in animal tissues and mixtures

Krzysztof Czamara, Agnieszka Kaczor

7.3.1. Classification of lipids

Lipids are undoubtedly one of the most essential biological molecules. Due to their structural diversity, directly related to their physical and chemical properties, lipids are the integral components of many cells and tissues. In biological systems, they perform a variety of functions from the cellular signaling through energy storage to the formation of biological membranes.

The term 'lipids' is derived from the Greek word λίπος (lípos), meaning fat, and refers to all greasy and fat-like compounds. According to a definition, a lipid is chemical compound which does not dissolve in water, but is soluble in solvents such as, for example petroleum ether, acetone or benzene [1].

However, rigorous treatment of this rule, based on the criterion of solubility, can lead to a misassignment or exclusion of certain compounds from the lipid group e.g. sphingomyelin, which is insoluble in diethyl ether. Another determinant used to classify a compound as a lipid is presence of the fatty acid residues in the structure and its natural origin. Currently, in times of highly developed organic synthesis, the latter criterion lost its applicability.

Due to the ambiguous definition lipids have the most comprehensive classification of all biomolecules. The most convenient classification scheme was proposed by the American biochemist Bloor in 1925 [2]. He divided all known (at the time) lipids into three groups: simple lipids, compound lipids and derived lipids (Fig. 7.3.1).

LIPIDS

Simple	Compound	Derived

Simple
- Neutral fats (acylglycerols)
- Waxes
  - True waxes
  - Cholesterol esters
  - Vitamin A and D esters

Compound
- Phospholipids
  - Lecithins
  - Cephalins
  - Sphingolipids
  - Phosphatidic acids
- Cerebrosides
  - Galactolipids
  - Glucolipids
- Sulfolipids

Derived
- Fatty acids
- Alcohols
  - Stright-chain alcohols
  - Sterols
  - Alcohols containing the β-ionone ring
- Hydrocarbons
  - Aliphatic hydrocarbons
  - Carotenoids
  - Squalene
- Vitamins D, E and K

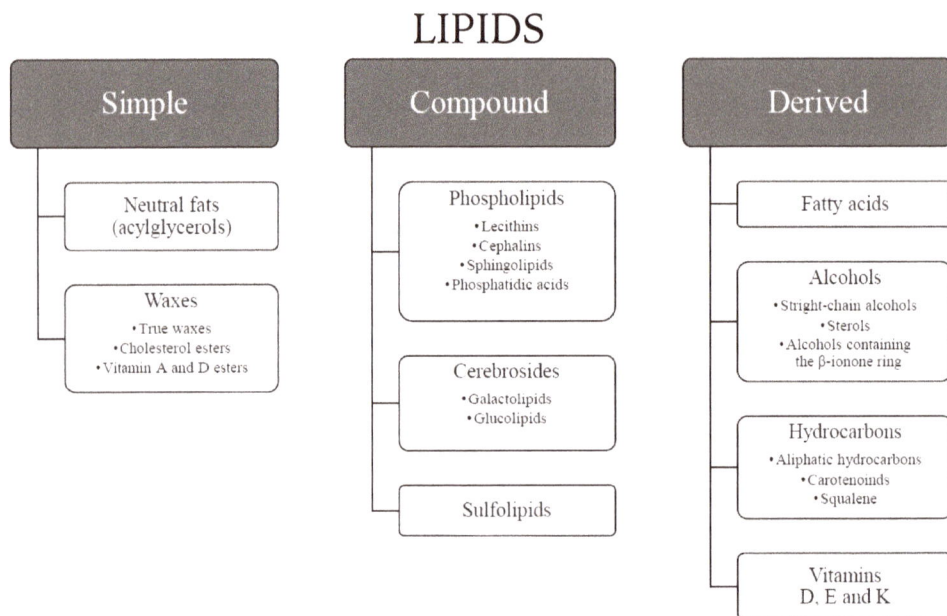

Fig. 7.3.1. The classification of lipids by Bloor [2,3]

Simple lipids are esters of fatty acids with various alcohols divided into two subgroups. The first is called neutral fats and contains acylglycerols, a large class of compounds due to the possibility of attachment one to three different fatty acid residues to the glycerol backbone. The second one are waxes that are widely distributed in plants and animals. This subgroup includes true waxes, the esters of fatty acids with long-chain alcohols, and alcohol esters of particular importance for human, among others cholesterol and vitamin A.

Compound lipids are divided into three classes due to the presence of not only fatty acid and the alcohol moieties, but an additional functional group in their structure. There are phospholipids, containing a phosphoric acid residue, cerebrosides with the sugar moiety instead of a simple alcohol and sulfolipids, derivatives of cerebrosides in which the C-3 carbon atom is bonded to a sulfuric acid residue. A subgroup of phospholipids is particularly interesting and important from the point of view of their functions. Lecithins and cephalins, belonging to phospholipids, are composed of glycerol, two fatty acid residues: saturated and unsaturated, a phosphoric acid residue and a nitrogen-containing base, for instance choline and ethanolamine occurring in lecithin and phosphatidylethanolamine, respectively. Sphingolipids, the characteristic lipids containing the amino alcohol – sphingosine, present in membranes of nerve cells, are also members of this subgroup.

The third group according to the Bloor's classification is group of derived lipids, *i.e.* all the chemical compounds formed by the hydrolysis of lipids and at the same time exhibiting characteristic lipids properties, e.g. solubility. There are four main sub-groups here. The most important is the first one constituting the basic building block of each lipid – a fatty acid. The multiple combinations of fatty acids with the compounds of other sub-groups: alcohols, sterols, carotenoids and vitamins results in a rich variety of naturally occurring lipids.

Since the development of the Bloor's classification scientists discovered a lot of new lipids, including synthetic, hence currently the group if lipids includes a vast amount of compounds. There is a lot of information about the structure, function and biochemistry for each group of lipids in the literature [4,5]. Studies on the properties of the lipids for many years lagged compared to other biomolecules *i.e.* proteins. The problem resulted primarily from the problem of isolation and separation of these compounds in a pure form, because most of them do not create crystals and occur as the amorphous mass. Only quite recently, the use of chromatographic and spectroscopic methods enabled to identify and classify lipids.

7.3.2. Spectroscopic characteristics of lipids

The knowledge of the chemical structure of a studied lipid enables efficient assignment of bands in the Raman spectrum. Therefore, it is essential to find out the positions of the marker bands and assign them to the vibrations of functional groups present in the lipid. Figure 7.3.2 shows Raman spectra of four standard members of different classes of lipids: fatty acids, triacylglycerols, cholesteryl esters and

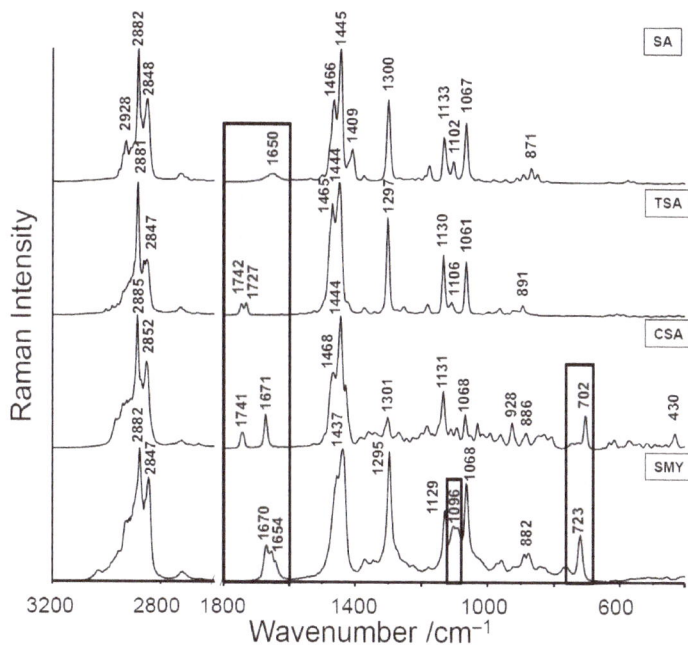

Fig. 7.3.2. The comparison of the Raman spectra of stearic acid (SA), glycerol tristearate (TSA), cholesterol stearate (CSA) and sphingomyelin (SMY). Selected fragments indicate the marker bands ranges for chosen compounds

sphingolipids. The most characteristic Raman bands in the spectrum of lipids are due to the presence of long fatty acid chains in their structures and are observed in the range of 1500-1400 cm^{-1}, 1300-1250 cm^{-1} and 1200-1050 cm^{-1}. These bands are attributed to scissoring and twisting vibrations of the CH_2 and CH_3 groups and C-C stretching vibrations, respectively. Additionally, a strong bands derived from the CH stretching vibrations are present in the high range of the spectrum (3100-2800 cm^{-1}). The other visible bands in the Raman profile of lipids related to the vibrations of characteristic functional groups are marked by the green rectangles indicating the wavenumber ranges enabling to distinguish and classify studied compounds.

Characteristic bands corresponding to the C=O and C=C stretching vibrations are observed in the spectral range of 1800-1600 cm^{-1}. Based on their location it is possible to identify whether the compound is a cholesteryl ester or acylglycerol and specify the type of geometric isomerism of bonds in the aliphatic chain. More-over, characteristic features occur in the Raman profile of some groups of lipids *i.e.* for the derivatives of cholesterol there is a band at *ca.* 700 cm^{-1}, resulting from the deformations of the sterol ring, and two typical bands at 1096 cm^{-1} and in the range of 720-780 cm^{-1} are observed for phospholipids, due to the presence of phos-phoric acid and nitrogenous base in their structure. The Raman spectrum of the sphingomyelin shows the band at 723 cm^{-1}, assigned to the stretching vibrations of the $N^+(CH_3)_3$ moiety.

The knowledge of the specific lipid marker bands is extremely useful for studying of more complex systems. The knowledge about the Raman spectra of pure substan-ces enables their identification in mixtures, cells or tissues [6,7]. An example of the use of Raman imaging for analysis of a lipid mixture is presented in Figure 7.3.3.

Based on the Raman spectra of pure compounds, marker bands were designated and used to create the Raman images of distribution for components in the mix-ture (Fig. 7.3.3b and c). Such analysis enables to determine the distribution, shape and size of the crystals of examined lipids in the complex samples.

AIM OF THE EXPERIMENT

1. Examination of the Raman profile of lipids belonging to different groups in order to define marker bands for their identification.
2. Application of the knowledge gained in part a) for analysis of lipids in the ani-mal tissue and/or lipid mixture.

SCIENTIFIC BACKGROUND

1. Fundamentals of Raman spectroscopy and Raman imaging; Chapters 2 and 4.4.
2. Construction and principle of operation of the dispersive Raman spectrometer; Chapter 2.
3. Basic knowledge about lipids and fatty acids; Chapter 7.2.

EQUIPMENT, MATERIALS, CHEMICALS

1. Confocal Raman spectrometer WITec equipped with laser with the excitation of 532 nm.

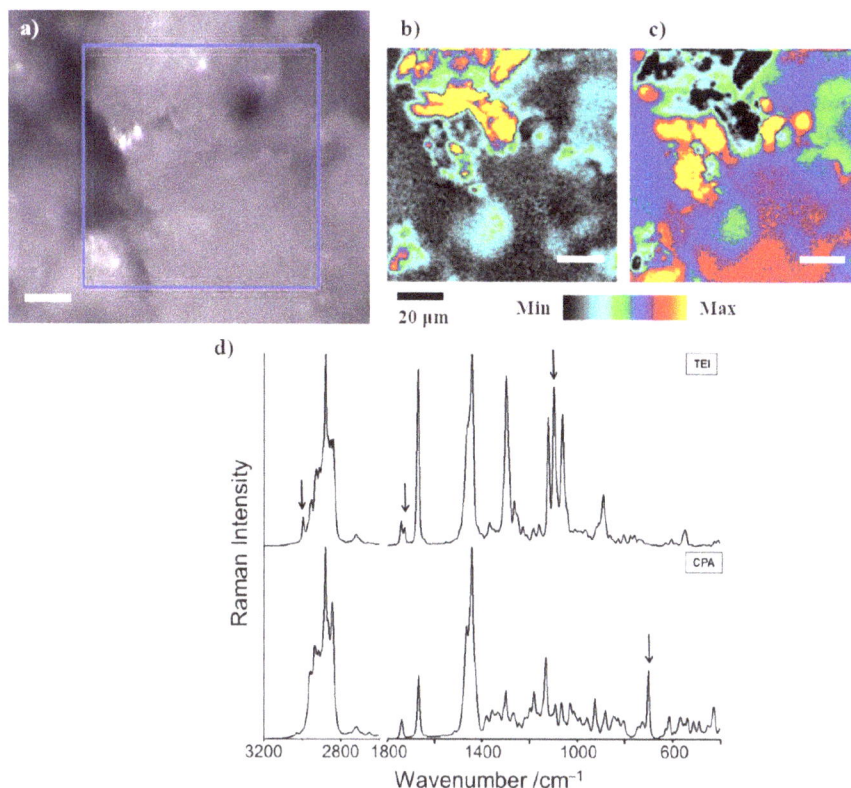

Fig. 7.3.3. The microphotograph (A) of a mixture of glycerol trielaidate (TEI) and cholesterol palmitate (CPA) with the Raman images of distribution of these components obtained by integration of bands at 2996 cm^{-1} (B) and 702 cm^{-1} (C). Raman spectra of the studied compounds (D) with indicated chosen marker bands (arrows)

2. Analytical standards of lipids belonging to different classes *i.e.* fatty acids, phospholipids, triacylglycerols, vitamins, cholesterol esters.
3. Fixed animal tissue *i.e.* mice aorta cross-section and/or mixture of 2-4 various lipids.
4. CaF$_2$ slides for measurements.

PROCEDURE

I. Examination of pure lipid standards
   1. Place the sample of a lipid on the CaF$_2$ slide.
   2. Place the slide on the scan table.
   3. Optimize the measurement conditions (focus, integration time, the number of scans, laser power) and collect single Raman spectrum. Proceed in the same way with the other samples.
II. Examination of lipids in complex samples
   1. Place the sample of a tissue/mixture on the CaF$_2$ slide.
   2. Place the slide the on scan table.

3. Optimize the measurement conditions (focus, scanning area, integration time, laser power), choose the number of points in the x and y axes and the size of the measured area and record a set of Raman spectra.
4. Perform the analysis of marker analysis. Prepare Raman images of distribution of individual lipids.

REPORT

1. Present the Raman spectra for all measured lipid standards in the 3200-400 cm^{-1} spectral range.
2. Assign the bands to vibrations on the basis of own knowledge and scientific literature.
3. Identify the similarities and differences in the spectra for different standards, discuss the results referring to the structural differences between compounds.
4. Denote the marker bands for each group of lipids. In the case of several lipids of the same group also indicate the bands enabling distinguishing of lipids.
5. Present the Raman images of distribution of lipids in a tissue/mixture obtained by the analysis of marker bands. Identify these lipids and compare your guess with literature spectra.
6. Comment on the usefulness of Raman spectroscopy to study lipid components in mixtures and tissues.

Refeences

1. Fahy E., Subramaniam S., Brown H.A., Glass C.K., Merrill A.H., Murphy R.C, et al. *A comprehensive classification system for lipids,* J. Lipid Res. **46**, 839 (2005).
2. Bloor W.R., *Biochemistry of the Fats,* Chem. Rev., 2, 243.
3. Deuel H.J. in: *The Lipids Their Chemistry and Biochemistry* (ed. H.J. Deuel) Interscience Publishers INC, New York 1951, pp. 1–6.
4. Gunstone F.D., Harwood J.L., in: *The Lipid Handbook 3rd Edition* (eds. F.D. Gunstone, J.L. Harwood, A.J. Dijkstra) CRC Press Taylor & Francis Group USA 2007, pp. 37–143.
5. Vance D.E, Vance J.E., *Biochemistry of lipids, lipoproteins and membranes, 5th edition,* Elsevier, The Nederlands 2008.
6. Pilarczyk M., Mateuszuk L., Ryguła A., Kepczynski M., Chlopicki S., Barańska M., Kaczor A., *Endothelium in Spots – High-Content Imaging of Lipid Rafts Clusters in db/db Mice,* PLoS One, **9**, e106065 (2014).
7. Majzner K., Kochan K., Kachamakova-Trojanowska N., Maslak E., Chlopicki S., Barańska M., *Raman imaging providing insights into chemical composition of lipid droplets of different size and origin: in hepatocytes and endothelium,* Anal. Chem., **86**, 6666 (2014).

7.4. Polymorphism of model triacylgliceroles

Marta Z. Pacia, Krzysztof Czamara, Agnieszka Kaczor

7.4.1. Polymorphism of lipids

Polymorphism is a characteristic phenomenon for fatty acids (FAs), as well as triacylglycerols (TAGs). It is defined as the existence of two or more forms or crystal structures for the same compound depending on the temperature, pressure and sample preparation. The existence of multiple polymorphs for FAs and TAGs is related to a different packing of the hydrocarbon chains [1]. For TAGs usually three different polymorphic forms: α, β' and β can be distinguished, but there are a few exceptions *i.e.* triheptadecanoin (trimargarin) that exhibits two modifications of the β' form (β_1' and β_2') [2] or trielaidin that exists solely as α and β polymorphs [3]. Each polymorphic state can be obtained upon defined crystallization conditions and thermal history. The α form, characterized by the lowest order and the smallest stability, is obtained by rapid cooling of the liquid triacylglycerol. This form crystallizes in the hexagonal system (H) of the smallest molecular packing [4]. The β' polymorph is acquired by gentle heating of the α form. Alkyl chains in the β' form are more closely packed compared to the a and organized in the orthorhombic perpendicular subcell ($O_\perp$). The most stable β form has the highest melting point and the tightest packing of the molecules in the triclinic ($T_\parallel$) crystallographic arrangement [5]. Thus, transitions from α to β *via* β' form undergoes during fast cooling of the liquid TAGs. The brief characteristics of triacylglycerols polymorphs is given in Table 7.4.1.

Molecules of individual polymorphs of TAGs are characterized by various arrangements of the carbon chains. They differ, in both, the lateral chain packing, determining the crystallographic structure of a polymorph (Fig. 7.4.1a) and lamella layer thickness (*d* spacing) (Fig. 7.4.1b). The *d* spacing depends on the length of molecules and the tilt angle between the chain axis and the lamellar plane. Subsequently in the α, β' and β forms the tilt angle decreases, resulting in the decrease of the *d* spacing and increase in the density of the lipid layer [6]. The arrangement of TAG molecules in different polymorphic phases is shown below (Fig. 7.4.1).

The polymorphism is crucial for industrial engineering related to the processing of fats and food production as it influences suitability of fats for consumption. For

Table 7.4.1. The characteristics of polymorphs of triacylglycerols

Property	α form	β' form	β form
Preparation	fast liquid cooling	heating of α form	heating of β' form or crystallization from solution
Stability	lowest	intermediate	highest
Subcell	hexagonal (H)	orthorhombic ($O_\perp$)	triclinic ($T_\parallel$)
Melting point	lowest	intermediate	highest

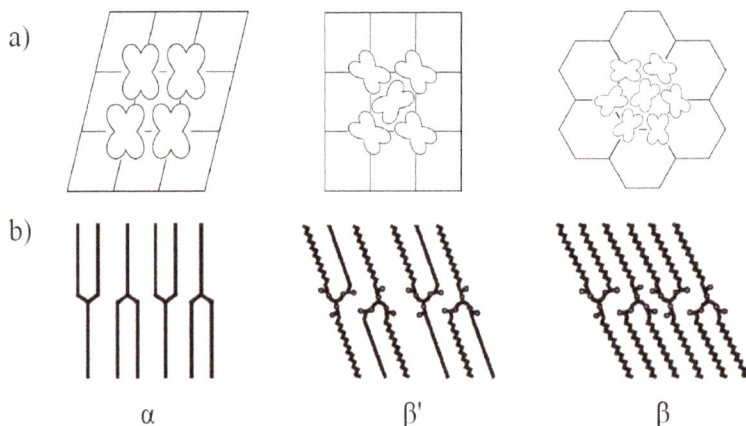

Fig. 7.4.1. The organization of aliphatic chains (A) and comparison of arrangement of TAG molecules (B) in α, β' and β forms

example, the manufacturing of chocolate requires strict compliance with the conditions of production due to the high content of cocoa butter present in five polymorphic forms [7]. Gourmets only accept chocolate in its crystal form V which has the noble surface sheen, crisp hardness and induces the pleasant melting sensation in the mouth [8]. Thus, the crystallization process has to be controlled by a sophisticated temperature regime called tempering, in order to make the chocolate crystallize exclusively in the form V [9].

7.4.2. Spectroscopic characteristics of TAGs polymorphs

Raman spectroscopy is a convenient method to investigate the polymorphism of TAGs. The structure of TAGs polymorphs have been previously studied in the context of the molecular organization, thermodynamics of phase transition and temperature dependence.

Below, tripalmitin (glyceryl tripalmitate, TPA) example is used to present the spectroscopic characteristics of TAGs polymorphs. As most triacylglycerols, TPA can exist in three polymorphic forms: α, β' and β. The α form is obtained by rapid cooling of the liquid compound to at least 0°C, hence further heating of the sample (α form) to the temperature of *ca.* 50°C to yield β' form. The β polymorph is formed by prolonged storage of the β' polymorph at the room temperature [10].

Raman spectroscopy enables clear discrimination of polymorphs. The influence of polymorphism on the Raman spectra for TPA is illustrated in Fig. 7.4.2.

The spectral range from 3200 to 600 cm⁻¹ is the most useful to evaluate spectroscopic changes among polymorphs. The high-wavenumber spectral region is helpful to determine polymorphs' order. Although the position of bands corresponding to the C-H stretching modes does not itself reveal significant differences between polymorphs, the ratio of the integral intensity of the bands at *ca.* 2850 to 2890 cm⁻¹

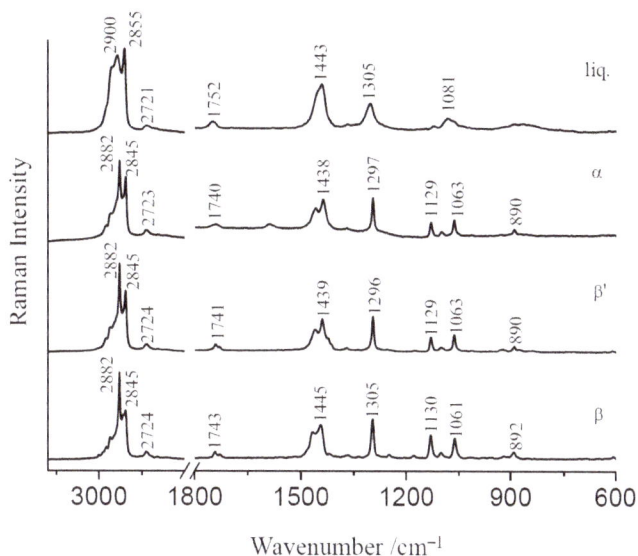

Fig. 7.4.2. Raman spectra of tripalmitin polymorphs and the liquid phase acquired with the 1064 nm laser line. (Reprinted (adapted) with permission from M. Pilarczyk, T.P. Wróbel, M. Barańska, A. Kaczor, J. Raman Spectrosc. 2012, **43**, 1515, Copyright © 2012 John Wiley & Sons, Ltd.)

($\eta = I_{2850}/I_{2890}$) can be applied to distinguish TAGs polymorphs (without analysis of the other spectral regions). This ratio, called η, decreases with the increase of thermodynamic stability. Additionally, three other spectral regions are of high importance for triacylglycerols' polymorphs recognition: 1770–1710 and 1500–1400 and 1140–1020 cm^{-1}, respectively (Fig. 7.4.3).

The analysis of the C=O stretching vibration range (1770–1710 cm^{-1}) enables discrimination of the β and β' polymorphs, as two counterparts of the ν(C=O) band are present in the β and β' polymorphs spectra contrarily to the isotropic

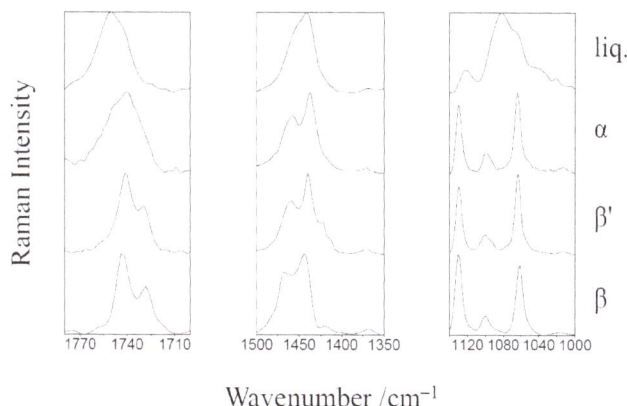

Fig. 7.4.3. Raman spectra of tripalmitin polymorphs and the liquid phase acquired with the 1064 nm in the key spectral ranges: 1780–1700 cm^{-1} (the C=O stretching vibrations region), 1500–1350 cm^{-1} (the CH$_2$ wagging vibrations region) and 1140–1000 cm^{-1} (the C-C stretching vibrations region)

α form and the liquid, where the single featureless band is observed (in the liquid at somehow higher wavenumber). Additionally, the β polymorph can be distinguished from the β′ form and the isotropic α polymorph based on the shape of the Raman features in the 1500–1400 cm^{-1} range. Depending on the conditions, the Raman spectra of solid *vs* liquid TAGs can significantly differ, especially in the 1140–1020 cm^{-1} range, exhibiting bands due to vibrations of the C-C bonds in hydrocarbon chains. For unsaturated triacylglycerols the increase of intensity of the middle band with increased the stability of the polymorph is observed, while for saturated TAGs this relationship is not obvious [11].

Fats polymorphism has to be taken into account when biological samples are analyzed as biological environments can stabilize certain structural lipid arrangements resulting in quite diverse Raman spectra.

AIM OF THE EXPERIMENT

The application of Raman spectroscopy to detection of polymorphic forms of triolein by the analysis of their Raman spectra.

SCIENTIFIC BACKGROUND

1. Fundamentals of Raman spectroscopy; Chapter 2.
2. Construction and principle of operation of the FT-Raman spectrometer; Chapter 2.
3. Basic knowledge about polymorphism of triacylglycerols.
4. Spectral characteristics of polymorphic forms of triacylglycerols.

FURTHER READING

1. Pilarczyk M., Wróbel T.P., Barańska M., Kaczor A., *Correlation of monomer structures of tripalmitin with the spectroscopic fingerprint of polymorphs: infrared, Raman and DFT study*, J. Raman Spectrosc. **43**, 1515 (2012)
2. Bresson S., El Marssi M., Khelifa B., *Conformational influences of the polymorphic forms on the CO and C-H stretching modes of five saturated monoacid triglycerides studied by Raman spectroscopy at various temperatures*, Vib. Spectrosc. **40**, 263 (2006)
3. Da Silva E., Bresson S., Rousseau D., *Characterization of the three major polymorphic forms and liquid state of tristearin by Raman spectroscopy*, Chem. Phys. Lipids., **157**, 113 (2009).
4. Czamara K., Majzner K., Pacia M.Z., Kochan K., Kaczor A., Barańska M., *Raman spectroscopy of lipids: a review*, J. Raman Spectrosc. **46**, 4 (2015).

EQUIPMENT, MATERIALS, CHEMICALS

1. FT-Raman spectrometer coupled with laser Nd:YAG (1064 nm)
2. Linkam THMS600 temperature stage
3. Samples of triacylglycerols e.g. triolein
4. Metal rings for measurements

PROCEDURE

1. Place the sample of triolein in the metal ring.

2. Place the metal ring into the temperature stage in the measurement chamber of the FT-Raman spectrometer.
3. Optimize the measurement conditions (focus, spectral resolution, the number of scans, laser power).
4. Record the Raman spectrum of triolein at the room temperature.
5. Using the temperature stage, cool the sample of triolein to $-150°C$ with the rate of $30°C \cdot min^{-1}$. Record the Raman spectrum at this temperature.
6. Place a new sample of triolein in the ring metal, cool to $-40°C$, held 5 min and subsequently heat the sample to $-28°C$. Record the Raman spectrum at final temperature.
7. Repeat the procedure. Place a new sample of triolein in the ring metal, cool to $-40°C$, held 5 min and subsequently heat the sample to $0°C$. Record the Raman spectrum at final temperature.

REPORT

1. Present the Raman spectra for polymorphic forms in the 3500-500 cm^{-1} spectral range, denote and assign the lipid marker bands.
2. Identify and discuss the differences in the spectra of the polymorphs of triolein.
3. Present the Raman spectra of polymorphic forms in the key spectral regions.
4. Calculate η, *i.e.* the ratio of the intensity of the bands at *ca.* 2850 to 2890 cm^{-1} for the various polymorphic forms by simply dividing the heights of the respective bands maxima. Confront the results with the literature data. [1]
5. Discuss the usefulness of Raman spectroscopy to study the polymorphic forms of triacylglycerols, particularly from the point of view of practical application of this method for studying lipids.

References

1. Simpson T.D., *Solid phases of trimargarin: A comparison to tristearin*, J. Am. Oil Chem. Soc., **60**, 95 (1983).
2. Dohi K., Kaneko F., Kawaguchi T., *X-ray and vibrational spectroscopic study on polymorphism of trielaidin*, J. Cryst. Growth, **237-239**, 2227 (2002).
3. Da Silva E., Bresson S., Rousseau D., *Characterization of the three major polymorphic forms and liquid state of tristearin by Raman spectroscopy*, Chem. Phys. Lipids, **157**, 113 (2009).
4. Bresson S., El Marssi M., Khelifa B., *Raman spectroscopy investigation of various saturated monoacid triglycerides*, Chem. Phys. Lipids, **134**, 119 (2005).
5. Bresson S., El Marssi M., Khelifa B., *Conformational influences of the polymorphic forms on the CO and C-H stretching modes of five saturated monoacid triglycerides studied by Raman spectroscopy at various temperatures*, Vib. Spectrosc., **40**, 263 (2006).
6. Lawler P.J., Dimick P.S., w: *FOOD LIPIDS Chem. Nutr. Biotechnol.* (Eds.: C.C. Akoh, D.B. Min), CRC Press, USA 2001, pp. 254–266.
7. Talbot G., w: *Ind. Choc. Manuf. Use* (Ed.: S.T. Beckett), John Wiley & Sons 2011, pp. 720.
8. Roth K., *Chocolate – The Noblest Polymorphism II*, DOI: 10.1002/chemv.201000030 (Roth K., *Von Vollmilch bis Bitter, edelste Polymorphie*, Chem. Unserer Zeit, **39**, 416 (2005).

9. Bakalis S., Le Révérend B.J.D., Anwar N.Z.R., Fryer P.J., *Modelling crystal polymorphisms in chocolate processing*, Procedia Food Sci., **1**, 340 (2011).
10. Kellens M., Meeussen W., Riekel C., Reynaers H., *Crystallization and phase transition studies of tripalmitin*, Chem. Phys. Lipids, **52**, 79 (1990).
11. Akita C., Kawaguchi T., Kaneko F., *Structural study on polymorphism of cis-unsaturated triacylglycerol: Triolein*, J. Phys. Chem. B, **110**, 4346 (2006).

7.5. Identification of carotenoids in plants by means of Raman spectroscopy

Aleksandra Jaworska, Małgorzata Barańska

7.5.1. Structure, functions and occurrence of carotenoids

Carotenoids are aliphatic or cyclic-aliphatic compounds, soluble in fats. They are divided into carotenes and xanthophylls (oxygen-containing derivatives of carotenes). Carotenoids which are present in plants are mainly tetraterpenoids C40 constructed from eight C_5 isoprenoid units. They are characterized by the presence of a long central chain with a system of conjugated double bonds (polyene chain), which absorbs light (electron transition $\pi - \pi^*$). Therefore, carotenoids exhibit yellow, orange and red colour [1].

Carotenoids play many important biological functions in living organisms. They occur together with the chlorophyll in the chloroplasts (green plant tissue) and chromoplasts, giving attractive colour of flowers and fruits. People or animals cannot synthesize carotenoids, but they can get them with food and then modify. For example, β–carotene after digestion with food is converted into vitamin A. It has been reported that carotenoids play an important role in prevention against cancer and cardiovascular diseases as well as a degeneration of the eye fovea. This activity is related to their antioxidant properties. Their ability to capture the singlet oxygen is mediated through a system of conjugated double bonds present in the molecules. It was found that the carotenes having at least 9 double bonds have the most important preventive significance. Therefore, in the recent years, the research has been focused largely on lycopene, which has 11 conjugated C=C bonds [1].

One of the most popular sources of carotenoids, namely lycopene and β–carotene, are tomatoes. The fruit colour depends on the ratio between the content of these two carotenes. In the red tomatoes the amount of lycopene is higher than in orange tomatoes whereas we can find more β–carotene in the latter. In tomato paste there is also a high content of lycopene, while in orange carrot root – β–carotene. In saffron, a spice derived from the flower of saffron crocus (*Crocus sativus*), we can find many carotenoids, including zeaxanthin, lycopene, and various α- and β-carotenes. However, its golden yellow-orange colour is primarily the result of crocin, containing seven C=C bonds. Lycopene, β–carotene and crocetin have 11, 9 and 7 conjugated double C=C bonds in the central chain, respectively (Fig. 7.5.1). Table 7.5.1 shows examples of carotenoids present in the selected plants.

7.5.2. Raman bands of carotenoids

Although carotenoids are present in plants in ppm amounts, they can be analyzed using Raman spectroscopy [2]. In the NIR-FT-Raman spectra carotenoid bands are strongly enhanced by a pre-resonance effect, therefore this group of biocompounds can be easily identified. The Raman technique also allows avoiding the fluorescence of biological material, observed with the Vis-lasers [3]. Strong bands of carotenoids are assigned to the in-phase stretching vibration of the C=C bond (v_1) and of the C-C bond (v_2). These bands are observed in the 1500–1580 and 1150–1170 cm^{-1} spectral regions, respectively. In addition, there is a medium--intensity band around 1000–1020 cm^{-1} (v_3), attributed to the CH$_3$ group wagging vibrations. For the identification of carotenoids in plants, the strong v_1 marker band is usually selected, as its position strongly correlates with the length of the polyene chain [4]. It was found that the wavenumber of the v_1 band is down-shifted along with lengthing the central chain due to the phonon-electron coupling [5].

Raman spectra of plant samples containing carotenoids can be recorded for both fresh and dried material. Spectra of red fruits of tomato, carrot root and saffron are shown in Fig. 7.5.2. It is clear from the examination of the spectral profile of these plants that a linear relationship appears between the length of the polyene chain of the carotenoid and the position of v_1.

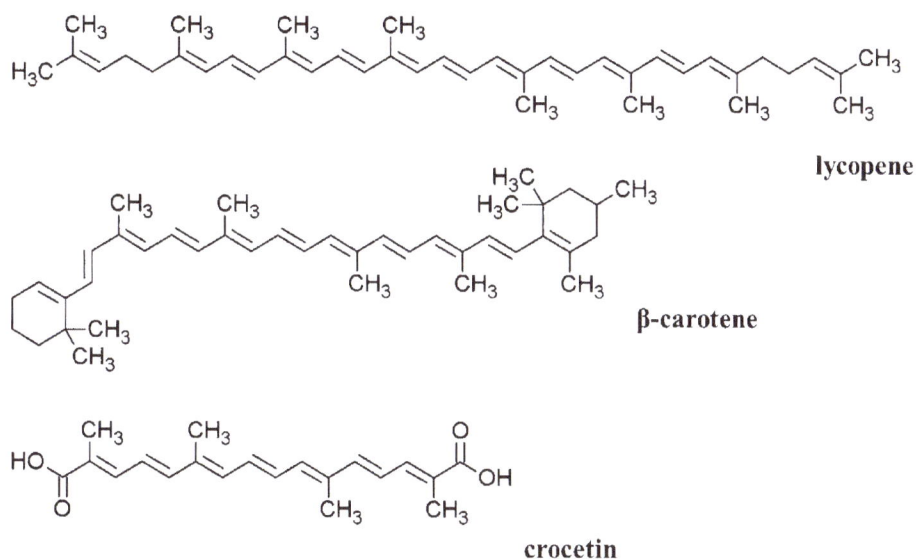

Fig. 7.5.1. The molecular structure of selected carotenoids

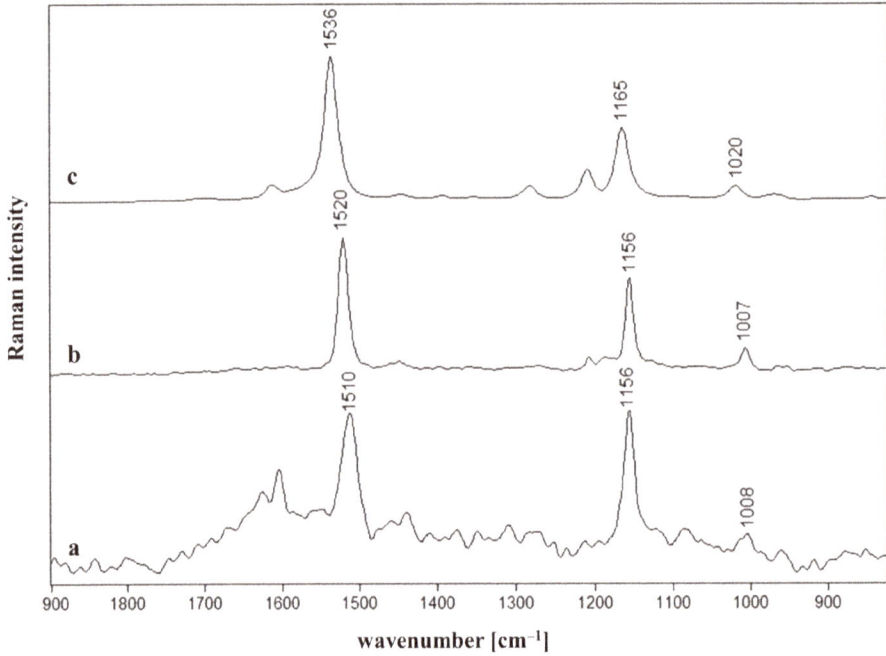

Fig. 7.5.2. Raman spectra of red tomato fruit (a), orange carrot root (b) and saffron (c)

Table 7.5.1. Predominant carotenoids in selected plants

Name of plant	Part of plant	Predominant carotenoid (number of C=C bonds in chain)
Saffron (*Crocus sativus* L.)	traits posts	crocetin (7)
Calendula (*Calendula officinalis* L.)	petal	auroxanthin (7), flavoxanthin (8), lutexanthin (8)
Chamomile (*Chamomilla recutita* L.)	pollen	flavoxanthin (8), lutexanthin (8)
Calendula (*Calendula officinalis* L.)	pollen	flavoxanthin (8), lutexanthin (8), lutein (9), antheraxsanthin (9)
Carrot (*Daucus carota* L.)	yellow root	lutein (9)
Carrot (*Daucus carota* L.)	leaf	lutein (9), β–carotene (9)
Ivy (*Hedera helix* L.)	leaf	lutein (9), β–carotene (9)
Basil (*Ocimum basilicum* L.)	leaf	lutein (9), β–carotene (9)
Begonia (*Begonia x semperflorens-cultorum* Hort.)	leaf	lutein (9), β–carotene (9)
Broccoli (*Brassica oleracea* var. *italica* L.)	flower	lutein (9), β–carotene (9)
Green beans (*Phaseolus vulgaris* L.)	green pod	lutein (9), β–carotene (9)
Maize (*Zea mays* L.)	seed	zeaxanthin (9)
Pumpkin (*Cucurbita pepo* L.)	fruit	β–carotene (9)
Carrot (*Daucus carota* L.)	orange root	β–carotene (9)

Table 7.5.1. (cd.)

Name of plant	Part of plant	Predominant carotenoid (number of C=C bonds in chain)
Pepper (*Capsicum annuum* L.)	green fruit	lutein (9)
Pepper (*Capsicum annuum* L.)	red fruit	capsanthin (9)
Watermelon (*Citrullus lanatus* Thumb.)	fruit	lycopene (11)
Tomator (*Lycopersicon esculentum* Mill.)	red fruit	lycopene (11)
Tomato (*Lycopersicon esculentum* Mill.)	orange fruit	β–carotene (9)
Tomato (*Lycopersicon esculentum* Mill.)	tomato paste	lycopene (11)

AIM OF THE EXPERIMENT

1. Detection of naturally occurring carotenoids by using Raman spectroscopy;
2. Correlation the chain length of carotenoids with bands in Raman spectrum;

SCIENTIFIC BACKGROUND

1. Theory of normal Raman effect; Chapter 2.
2. Construction of the FT-Raman and dispersive Raman spectrometers; Chapter 2.
3. Structure and occurrence of basic carotenoids.

EQUIPMENT, MATERIALS, CHEMICALS

1. Natural samples containing carotenoids such as carrot, tomato, orange, pepper, etc.
2. Fourier-transform Raman spectrometer (e.g. Bruker).
3. Software to analyse spectra, e.g. OPUS, Origin.

PROCEDURE

1. Determine, on the basis of the scientific literature, which carotenoids are present in samples and how long is their central chain.
2. Measure Raman spectra of samples brought by students, including optimizing measurement conditions (number of scans per sample, laser power) to obtain good signal to noise ratio in the shortest possible time in the spectral range from 0 to 4000 cm^{-1}.
3. Determine and indicate marker bands for carotenoids according to Fig. 7.5.2.

REPORT

1. Read the band positions in Raman spectra of carotenoids (v_1, v_2 and v_3).
2. Determine the length of the polyene chain of carotenoids on the basis of the position of the band v_1.
3. Show graphically the relationship between the value of v_1 and the $(N+1)^{-1}$, where N is a number of the conjugated C=C bonds in a carotenoid, and discuss

this relationship. Estimate the values of v_1 for the carotenoids containing 5, 13 and 17 C=C bonds in the central chain.

4. Discuss the usefulness of Raman spectroscopy in determination of carotenoids in biological samples.

References

1. Y. Ozaki et al., *Potential of Near-infrared Fourier Transform Raman spectroscopy in food analysis*, Appl. Spectrosc., **46**, 1503 (1992).
2. B. Schrader, *Infrared and Raman Spectroscopy. Methods and Applications*, VCH, Weinheim 1995.
3. H. Schulz et al., *Potential of NIR-FT-Raman Spectroscopy in Natural Carotenoids Analysis*, Biopolymers, **77**, 212 (2005).
4. R. Withnall et al., *Raman Spectra of Carotenoids in Natural Products*, Spectrochim. Acta A, **59**, 2207 (2003).
5. M.A. Strehle et al., *Controlling the Quality of Different Vegetable and Fish Oils by Means of Raman Spectroscopy*, Proceedings XIXth International Conference on Raman Spectroscopy, P.M. Fredericks, R.L. Frost, L. Rintoul (eds.), CSIRO Publishing 2004, 411-412.

7.6. Identification of terpenes in citrus oils by means of Raman spectroscopy

Aleksandra Jaworska, Małgorzata Barańska, Kamilla Malek

7.6.1. Essential oils – occurrence and composition

The oils of various citrus plants (e.g. orange, grapefruit, mandarin, lemon and lime) are frequently used as raw materials in the perfume, cosmetics and flavour industry [1]. Below there are brief descriptions of the aromatic properties of the selected citrus oils [2]:

Sweet Orange Essential Oil (Citrus sinensis)

Sweet Orange Essential Oil is widely available and is amongst the most inexpensive of all essential oils. It is becoming more popular and included in commercial cleaners as it can help to naturally cut grease.

Lemon Essential Oil (Citrus limon)

Lemon Essential Oil has a powerfully fresh traditional lemon fragrance that is quite energizing and uplifting. It is a good choice to diffuse it when trying to clear a room of the smell of cigarette smoke or other unpleasant aromas.

Lime Essential Oil (Citrus aurantifolia)

Cold pressed Lime Essential Oil is the most aromatically potent of the fruity citrus oils. It is well known in folklore for its ability to cleanse, purify and renew the spirit and the mind, as well as it is said to be effective in cleansing the aura. Lime Essential Oil is sometimes found steam distilled.

Grapefruit Essential Oil (Citrus paradisi)

Grapefruit Essential Oil is sweet-smelling, bright and uplifting. It is an energizing and uplifting oil, and it is wonderful to diffuse in the mornings or while working out to help to stay awaken and be energetic.

Mandarin Essential Oil (Citrus reticulata)

Mandarin Essential Oil is a favorite oil of children and parents. Of all the citrus oils, Mandarin Essential Oil is the sweetest and tends to be the most calming.

Many of the citruses oils are phototoxic. When exposed to sunlight, the naturally occurring chemical constituents found in some citrus essential oils become phototoxic. Bergamot, Bitter Orange, Lemon and Lime are amongst the citrus oils that are generally regarded to be highly phototoxic.

The essential oils, which consist mainly of terpene hydrocarbons (Fig. 7.6.1), are usually obtained by pressing the fruit peels. The main component of citrus oils is limonene (~45% in lime oil and up to ~96% in orange or grapefruit oil). Other terpenes, such as α– and β-pinene, γ-terpinene, terpinolene and sabinene, which are known to be very unstable in the presence of light and oxygen, can be also detected in the oils.

7.6.2. Identification of essential oils by means of gas chromatography and Raman spectroscopy

Gas chromatography has been widely applied to identify the individual citrus oil composition (see table 7.6.1). However, since these measurements are expensive and time-consuming, some attempts have been made to develop a Raman and IR

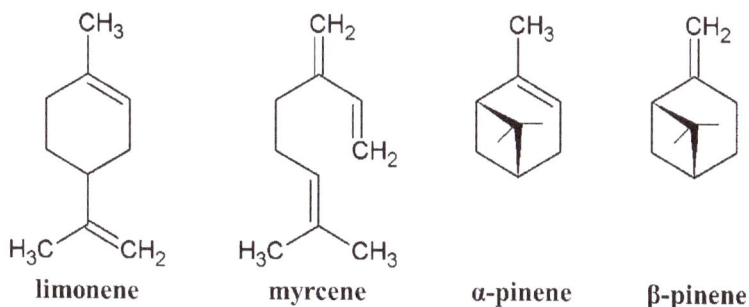

Fig. 7.6.1. Structures of basic terpenes

Table 7.6.1. Gas chromatographic composition [%] of the most common citrus oils [1]

Citrus oil	limonene	myrcene	α-pinene	β-pinene
grapefruit	94.8	1.8	–	–
sweet orange	95.1	1.8	0.6	–
mandarin	77.3	1.7	1.8	1.1
lemon	68.6	1.5	1.9	12.1
bitter orange	95.5	1.8	0.5	–
lime	49.9	1.3	1.4	4.2

Table 7.6.2. Assignments of main bands appearing in Raman spectra of terpenes [1,2]

Spectral range [cm^{-1}]	Assignment
645 – 760	ring vibration
~1293	C-O stretching
1424 – 1445	CH_2 scissoring
1634 – 1657	ring vibration
1672 – 1701	C=O stretching

methods for the characterization of various citrus oils. Table 7.6.2 collects the summary of positions of characteristic bands observed in the Raman spectra of the most popular terpenes [1] along with their assignment to vibrations of functional groups.

7.6.3. Chemometric analysis of Raman spectra of essential oils

Study of the composition of the essential oils and identification of main components can be performed by means of chemometric analysis. It allows grouping the spectra according to their spectral similarities, and therefore the chemical composition (Fig. 7.6.2.). It facilities data analysis, especially when marker bands of the main components (terpenes) are not clearly visible in the spectra.

AIM OF THE EXPERIMENT
Detection and classification of naturally occurring essential oils by using Raman spectroscopy

SCIENTIFIC BACKGROUND

1. Theory of normal Raman effect; Chapter 2.
2. Construction of the FT-Raman and dispersive Raman spectrometers; Chapter 2.
3. Chemical structure of terpenes and their occurrence in essential oils.

FURTHER READING

1. Schulz H., Schrader B., Quilitzsch R., Steuer B., *Quantitative analysis of various citrus oils by ATR/FT-IR and NIR-FT Raman spectroscopy*, Appl. Spectrosc., **56**, 117 (2002).

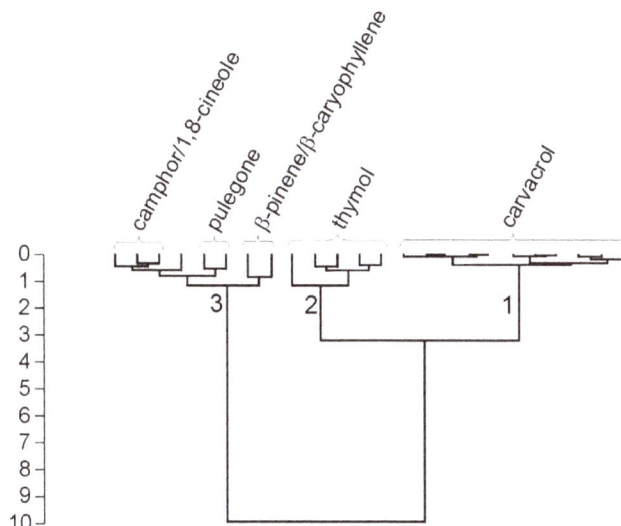

Fig. 7.6.2. Chemometric analysis (Ward's algorithm) of the Raman spectra of the aromatic plants collected in Turkey showing the main monoterpenes present in essential oils of these plants [2]

2. Schulz H., Özkan G., Barańska M., Krüger H., Özcan M., *Characterisation of essential oil plants from Turkey by IR and Raman spectroscopy*, Vib. Spectrosc., **39**, 249 (2005).

EQUIPMENT, MATERIALS, CHEMICALS

1. Samples of terpenes as reference substances (limonene, myrcene, α-pinene, β-pinene), extracted essential oils from lemon, orange, grapefruit, and fresh pills of citrus fruits (orange, citrus, grapefruit) brought by students.
2. Fourier transform Raman spectrometer.
3. Software for spectral analysis, e.g. OPUS.

PROCEDURE

1. Determine, on the basis of the scientific literature, which terpenes are present in samples and what are their structures.
2. Measure Raman spectra of pure terpenes as references, extracted essential oils and pills of citrus fruits, in the spectral range from 0 to 4000 cm^{-1}.
3. Determine marker bands of terpenes from spectra of references substances.
4. Identify marker bands of terpenes in Raman spectra of extracted essential oils and natural products.

REPORT

1. Propose the marker bands in Raman spectra of pure terpenes and assign vibrations to these bands according to Table 7.6.2.
2. Based on Table 7.6.2 identify terpenes present in the citrus oils and fruit pills.
3. Indicate marker bands present in the measured Raman spectra of oils and peels, and assign them to the reference substances.

4. Correlate terpenes identified in Raman spectra of the essential oils and fruits pills with the composition determined with the use of gas chromatography (Table 7.6.1).
5. Perform chemometric analysis of spectra of essential oils due to group the spectra according to the chemical composition. Identify the main components of measured oils.
6. Discuss the usefulness of Raman spectroscopy in detection of terpenes in natural products.

7.7. An analysis of pigments and painting materials in Raman spectra

Anna Ryguła, Kamilla Malek

There is no doubt that the cultural heritage of both the oldest, from archaeological sites as well as those that are witness to the events of the last few centuries, being a work of art or just the subject of daily use requires a proper care. It involves using the appropriate conditions of storage of art objects in museums as well as applying appropriate methods of conservation to stop gradual deterioration or restore colours and forms. Objects of conservators' interest are paintings, ceramics, textiles, glass, furniture, statues, books or plastics. They consist of both organic and inorganic materials of a natural origin (plant or animal sources) or of a synthetic origin. The key issue in conservation works is to identify and understand the mechanisms of degradation occurring in works of art. This information is necessary to stop further destruction process and to take treatments such as cleaning or consolidation of structure. A considerable issue is also an opportunity to determine the origin of the object through the identification of chemical composition. This not only highlights the realities of the times or the workshop of the old masters, but also confirms the authenticity of works of art.

7.7.1. Raman spectroscopy as an analytic technique in conservation of art works

Today, many analytical techniques, both physical and chemical, are employed in the wide field of conservation. Investigations regard elemental composition as well as physical properties of the object. However conservation methods must have one important feature – the destruction of the work of art during the analysis must be minimized, so techniques in which a minimum sample is taken or even non-invasive methods are highly desirable.

The development of spectroscopic techniques, in particular Raman spectroscopy over the past 20 years has resulted in new possibilities of use of these methods in conservation research. These methods are used because of their simplicity, data collection *in situ* and because these methods are non-destructive and does not require pre-treatment of the sample. The progress of technology and miniaturization of instrumentation allows for the Raman spectroscopic analysis without the need to visit a laboratory, which is possible by using a portable spectrometer, see Figure 7.7.1A [1–3]. The use of a Raman microscope also enables an accurate analysis on a micro scale, e.g. grains of pigment mixture or a complex sample surface [1,4,5] as illustrated in Figure 7.7.1 B.

Chemical and conservation reports, which have appeared in the literature, have shown a growing interest in the application of Raman spectroscopy in the analysis of works of art. Published research have focused on the analysis of paint pigments detected in manuscripts, murals, paintings as well as lithographs, ceramics, glass and metal. At the same time Raman studies on organic materials such as textiles, leather, rubbers, resins, waxes, wood or organic dyes have been carried out. The popularity of spectroscopic methods are evidenced by the creation of databases containing standard spectra substances, especially inorganic compounds, present in objects undergoing conservation [1-9].

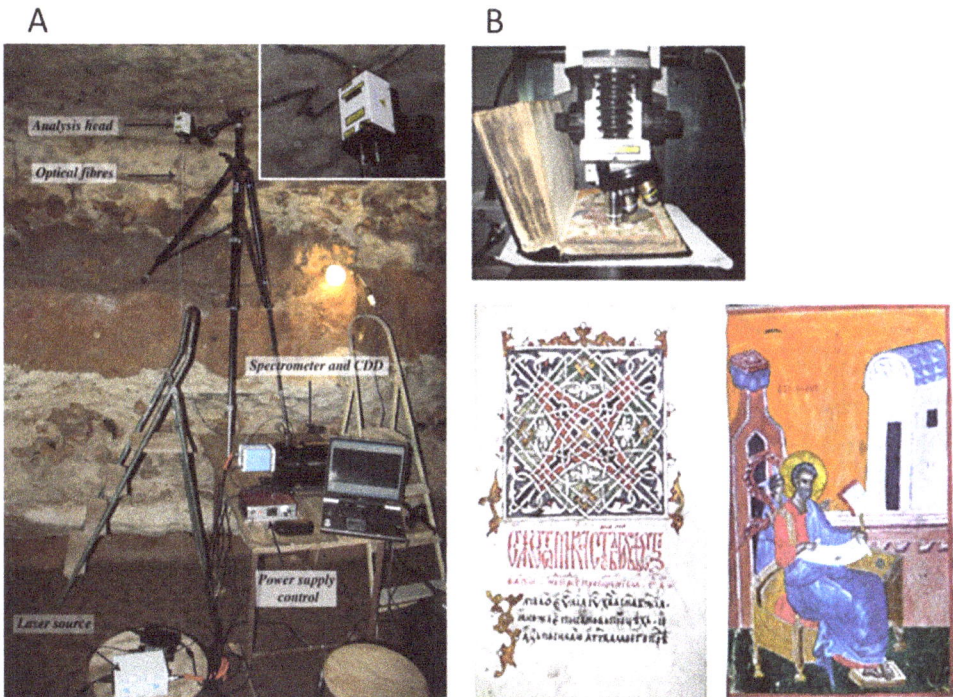

Fig. 7.7.1. *In situ* Raman measurements: (A) analysis of prehistoric drawings found in the cave of Rouffgnac-Saint-Cernin (Dordogne, France) [3]; (B) analysis of inks and pigments in illuminated medieval old-Slavonic manuscripts [4].

7.7.2. Structure of painting layers

Before examining paint layers one should first consider the type of a material, which will be identified. The two most important components of paint are coloured substance (a pigment or a dye) and a binder. Apart from them, artists apply various kinds of additives and fillers [4].

There are two types of coloured substances, pigments and dyes. Pigments show a varying degree of granularity. They are coloured solids insoluble in a binder and forming a suspension. They are divided into two classes, inorganic and organic. The former includes mineral paints. A dye at the other hand, is a substance soluble in fluid, and it interacts directly with the surface and staining it, like in fibrous. Natural dyes were usually prepared from plants or animals [4,10].

The difficulty in the analysis of natural inorganic pigments is their different chemical composition or different stoichiometry depending on the origin on contrary to synthetically prepared pigments. For example, verdigris has chemical formula $x\mathrm{Cu(CH_3COO)_2} \cdot y\mathrm{Cu(OH)_2} \cdot z\mathrm{H_2O}$, where x, y and z are variables, while the Egyptian green and blue differ only in the ratio of metal ion to silicates ($\sim\mathrm{CaCuSi_4O_{10}}$). Some synthetic pigments were already synthesized in antiquity, e.g. lead white, Egyptian blue. Originally organic pigments were only extracted from plants or animals. Since the late XIXth century, this group of pigments is obtained synthetically.

Fillers are inert substances, mostly transparent or of white colour, added to pigments to improve their properties or increasing volume, thereby reducing the cost of paint. Common fillers are chalk (calcium carbonate), gypsum (hydrated or anhydrous calcium sulphate), kaolin (or clay – hydrated aluminium orthosilicates [4, 10]), barium white (a natural mineral less opaque than a synthetic counterpart – barium sulphate) or talc. Examples of coloured fillers are sifted ash and volcanic glass.

Binders are, in turn, liquid or semi-liquid substances, in which grains of pigments and fillers are suspended. Their role is to keep components of paint within a layer and the attachment of paint to the ground of a painting. Several organic compounds play a role of a binder such as oils, waxes, resins, synthetic adhesives, balsams, proteins, temperas [4,10].

7.7.3. Methodology of a Raman analysis of works of arts

All the components of painting layers give Raman spectra as pure substances, although cross section of Raman scattering phenomenon of minerals is much higher than compounds of an organic origin. Therefore, Raman spectra of the paint layer often exhibit only bands of minerals (pigments and fillers).

Raman spectroscopy is sensitive not only to a chemical but also to crystal structure. An example is lead oxide (II) (PbO), *i.e.* yellow massicot with rhombic crystalline structure, red-orange tetragonal litharge and red lead ($\mathrm{Pb_3O_4}$) [7]. Raman spectra of these compounds are independent of a laser excitation line and show the sensitivity of this technique to vibrations of the functional group as well as crystal lattice. Despite the fact that massicot and litharge are chemically identical, their different crystal structure provides specific spectral information.

Coupling of a Raman spectrometer with an optical microscope allows to record spectra from an area with a diameter of 0.3 μm. The size of the laser spot depends on optical components of the microscope and the laser wavelength. Generally, the longer the wavelength of laser excitation and a lower the magnification of the microscope objective, the smaller later resolution is achieved (see Chapter 4.4). For example, Raman imaging using a laser excitation at 1064 and 488 nm and an objective with magnification 40× provides a maximum lateral resolution of 0.99 and 0.46 μm, respectively, which obviously influences the quality of the chemical information identified in Raman spectra.

The identification of a painting material is based on the comparison of Raman spectrum with spectra available in databases [11-13]. There is no one complete database of materials used in works of art and archaeological objects, although a number of publications reporting Raman spectra of minerals, mineral pigments, historical and contemporary organic pigments, glazes and paint binders has increased. An example of such an analysis at the micro level is the identification of painting materials used by unknown author to create "Ecce Homo" (Fig. 7.7.2 A) [14]. A protocol of the investigations relied on the collection of microsamples during a conservation process, Fig. 7.7.2 B, from points marked in Fig. 7.7.2 A and then the collection of Raman imaging with using of 1064 and 532 nm laser (Fig. 7.7.2 C). Next the recorded Raman maps were analysed with chemometric tools and then spectra were compared with spectra of reference materials of a colour specific for the layer or the pigment grains (Fig. 7.7.2 C).

The Raman analysis is usually complemented by an elemental analysis using ESEM-EDS or infrared absorption spectroscopy that is a complementary technique to Raman spectroscopy (see Chapter 6.6).

One should also pay attention to the wavelength of a laser employed in a Raman spectrometer. The most common are lasers with an excitation in the visible region of radiation than those in near infrared. On the one hand it results from the access to instrumentation because dispersive spectrometers are widespread (see Chapter 2). On the other hand, taking into account the resonance Raman effect the Vis illumination can lead to interesting results. However, the Raman analysis entails some problems, mainly due to fluorescence phenomenon occurring in Vis region. This can be avoided by changing the excitation range to near infrared [1,2,12,15,16]. In the case of the painting samples, the problem of fluorescence can be caused by a dye and a binding medium or by grounding and impurities such as fats and soot. Painting layers are also very sensitive to localized heating of the sample with the laser light, which often leads to thermal degradation of the sample. In addition, a quantitative analysis of components of paint layers is rather impossible, e.g. the determination of the concentration of each pigment or pigment to binder ratios.

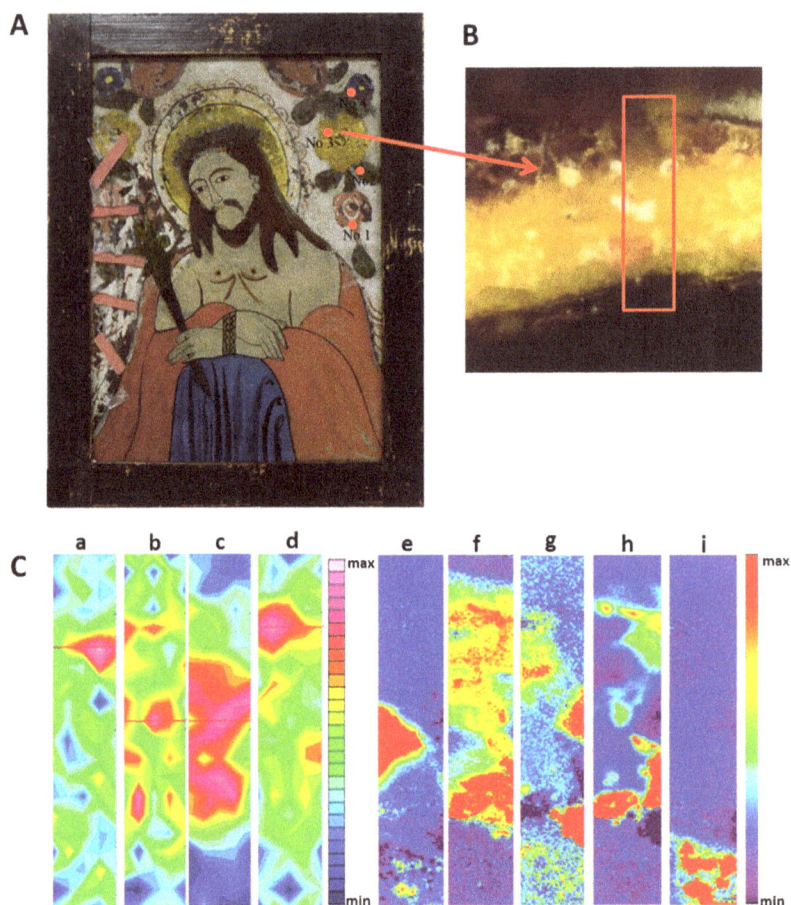

Fig 7.7.2 (B) Microphotography of a cross section collected from a glass painting „Ecce Homo" (an unknown author) (A) imaged by using a FT-Raman and dispersive microscopes (excitation lasers at 1064 and 532 nm, respectively). (C) Raman images (laser: 1064 nm), based on intensities of bands at: a) 122cm^{-1} (red lead), b) 838 cm^{-1} (chrome yellow), c) 988 cm^{-1} (barium sulphate), d) 1008 cm^{-1} (gypsum); Raman images (laser: 532 nm), based on intensities of bands at: e) 139 cm^{-1} (litharge), f) 831 cm^{-1} (chrome yellow), g) 1008 cm^{-1} (gypsum), h) 979 cm^{-1} (barium sulphate), i) 1087 cm^{-1} (calcite) [14]

AIM OF THE EXPERIMENT

1. Creation a database of reference spectra and identification of painting materials by using Raman spectroscopy.
2. Collection of Raman spectra of painting layers and interpretation of results.
3. An understanding of advantages and limitations of Raman spectroscopy in the determination of chemical composition of a paint layer.

SCIENTIFIC BACKGROUND

1. Fundamentals of Raman spectroscopy and its imaging technique; Chapter 2 and 4.4.
2. Construction of a Raman spectrometer; Chapter 2.
3. Basic information about chemical composition of a paint layer.

FURTHER READING

1. Bell I.M., Clark R.J.H., Gibbs P.J., *Raman spectroscopic library of natural and synthetic pigments (pre- ~1850 AD)*, Spectrochim. Acta Part A, **53**, 2159 (1997).
2. Burio L., Clark R.J.H., *Library of FT-Raman spectra of pigments, minerals, pigment media and varnishes, and supplement to existing library of Raman spectra of pigments with visible excitation*, Spectrochim. Acta Part A, **57**, 1491 (2001).
3. Smith G.D., Clark R.J.H., *Raman microscopy in archaeological science*, J. Arch. Sci., **31**, 1137 (2004).

EQUIPMENT, MATERIALS, CHEMICALS

1. Selected pigments of white colour (e.g. zinc white, lead white, anatase), yellow colour (e.g. tin-lead yellow, massicot, chrome yellow, orpiment) and blue colour (e.g. smalt, azurite, Prussian blue, cobalt blue, ultramarine, indigo) or others; oil binder and selected fillers (e.g. chalk, blank fix, lithopone, gypsum); a cross-section of painting layers containing oil binder and yellow, blue and white pigments. Painting layers can be prepared by students during the exercise.
2. Raman spectrometer.
3. Software for spectral analysis, e.g. OPUS, Origin.

PROCEDURE

1. Measure Raman spectra of chosen pigments, oil binders and fillers in the form of a solid as reference substances.
2. Prepare a painting layer on an oil binder with different ratios of pigments and fillers.
3. Measure Raman spectra for painting layers of a known composition prepared during the practical and a sample of unknown composition provided by the assistant.

REPORT

1. Create a database of reference spectra for a selected group of painting (including their chemical composition, colour and period of application in painting) and determine their marker bands.
2. Present the Raman spectra of painting layers prepared by the students and in the sample of unknown composition and identify used materials. Analyse the results for the known composition of the paint layer and the sensitivity of Raman technique.
3. Discuss advantages and disadvantages of Raman spectroscopy for the detection of painting materials.

References

1. Bellot-Gurlet L., Pagés-Camagna S., Coupry C., *Raman spectroscopy in art and archaeology*, J. Raman Spectrosc., **37**, 962 (2006).

2. Vandenabeele P., Lambert K., Matthys S., Schudel W., Bergmans A., Moens L., *In situ analysis of mediaeval wall paintings: a challenge for mobile Raman spectroscopy*, Anal. Biol. Chem., **383**, 707 (2005).
3. Lahlil S., Lebon M., Beck L., Rousselière H., Vignaud C., Reiche I., Menu M., Paillet P., Plassard F., *The first in situ micro-Raman spectroscopic analysis of prehistoric cave art of Rouffignac St-Cernin, France*, J. Raman Spectrosc., **43**, 1637 (2012).
4. Nastova I., Grupče O., Minčeva-Šukarova B., Turan S., Yaygingol M., Ozcatal M., Martinovska V., Jakovlevska-Spirovska Z., *Micro-Raman spectroscopic analysis of inks and pigments in illuminated medieval old-Slavonic manuscripts*, J. Raman Spectrosc., **43**, 1729 (2012)
5. Clark R.J.H., *Raman microscopy: application to the identification of pigments on medieval manuscripts*, Chem. Soc. Rev., **24**, 187 (1995).
6. Staniszewska E., *IR and Raman imaging of cross-sections of a glass painting "Ecce Homo"* (in Polish), MSc thesis, WCh UJ, Krakow, 2011.
7. Smith G.D., Clark R.J.H., *Raman microscopy in archaeological science*, J. Arch. Sci., **31**, 1137 (2004).
8. Vandenabeele P., Edwards H.G., Moens L., *A decade of Raman spectroscopy in art and archaeology*, Chem. Rev., **107**, 675 (2007).
9. Vandenabeele P., Wehling B., Moens L., Edwards H., De Reu M., Van Hooydonk G., *Analysis with micro-Raman spectroscopy of natural organic binding media and varnishes used in art*, Anal. Chim. Acta, **407**, 261 (2000).
10. Hopliński J., *Paints and binding media* (in Polish), Wydawnictwo: Zakład Narodowy im. Ossolińskich, Wyd. II, Wroclaw 1990.
11. Castro K., Perez-Alonso M., Rodriguez-Laso M.D., Fernandez L.A., Madariaga J.M., *On-line FT-Raman and dispersive Raman spectra database of artists' materials (e-VISART database)*, Anal. Bioanal. Chem., **382**, 248 (2005).
12. Burio L., Clark R.J.H., *Library of FT-Raman spectra of pigments, minerals, pigment media and varnishes, and supplement to existing library of Raman spectra of pigments with visible excitation*, Spectrochim. Acta Part A, **57**, 1491 (2001).
13. Bell I.M., Clark R.J.H., Gibbs P.J., *Raman spectroscopic library of natural and synthetic pigments (pre- ~1850 AD)*, Spectrochim. Acta Part A, **53**, 2159 (1997).
14. Staniszewska E., Malek K., Kaszowska Z., *Studies on paint cross sections of a glass painting by using FT-IR and Raman microspectroscopy supported by univariate and hierarchical cluster analyses*, J. Raman Specrosc., **44**, 1144 (2013).
15. Schrader B., Schulz H., Andreev G.N., *Non-destructive NIR-FT-Raman spectroscopy of plant and animal tissues, of food and works of art*, Talanta, **53**, 35 (2000).
16. Andreev G.N., Schrader B., Schulz H., Fuchs R., Popov S., Handjieva N., *Non-destructive NIR-FT-Raman analyses in practice. Part I. Analyses of plants and historic textiles*, Fresenius J. Anal. Chem., **371**, 1009 (2001).

7.8. Detection and determination of glucose in pharmaceuticals and body fluids

Agnieszka Kaczor

Raman spectroscopy enables detection of a studied compound and analysis of its structure. In case if calibration of a method, quantitative assay of the compound in a studied sample is also possible. In double-beam FT-IR and UV-Vis spectrometers the band intensity is defined by Beer-Lambert-Bouguer's law stating that the band intensity, represented as an integral of the area of this band, is directly proportional to the concentration of the measured species and thickness of the sample. It means that for the samples of the same thickness and equal concentration of a studied compound, but analyzed on different FT-IR or UV-Vis spectrometers, bands intensity is exactly repeatable if the same basic measurement parameters such as spatial resolution, apodization, etc. are applied. Therefore, in these cases, the band intensity is a function of a sample independent on other measurement parameters and the ordinate is described in the absolute units of absorbance/transmittance. In Raman spectroscopy the number of scattered photons, dependent on various factors (the optical pathway, analyte concentration, integration time, power and wavenumber of the excitation light, etc.), is counted, so, the intensity of scattered light is expressed in relative units. A scattered intensity depends, among others, on the laser power, excitation wavelength, sample orientation and, as such, is not a simple function of concentration and sample thickness. Additionally, in dispersive Raman spectrometry, the band intensity depends on the integration time. Therefore, application of Raman scattering spectroscopy in quantitative analysis is difficult and requires calibration and unknown sample measurements to be done on the same instrument and careful optimization of parameters (laser power, focusing, spectral resolution, sample orientation, radiation pathway). Under these circumstances the band intensity becomes directly proportional to the concentration of the analyte.

Quantitative analysis with the application of Raman spectroscopy are usually conducted in the solution as effects due to size and shape of surface are in this case eliminated. The very important advantage of Raman spectroscopy is possibility of working in aqueous solutions as water is a very poor Raman scatterer. It enables measurements of various bioactive compounds such as drugs and their metabolites in body fluids.

Diabetes mellitus

Glucose is a monosaccharide (simple sugar) containing six carbon atoms in a molecular (hexose) and the aldehyde functional group (aldose). The structural formula of D-glucose is presented in Fig. 7.8.1.

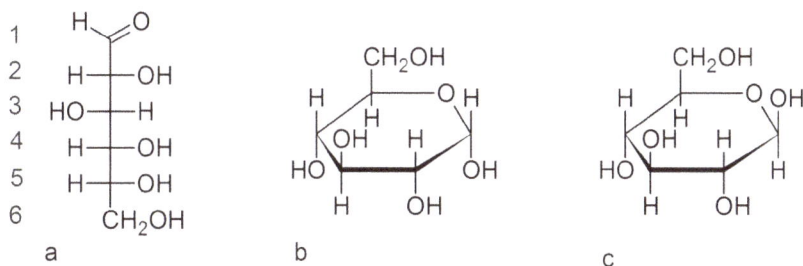

Fig. 7.8.1. Structural formulas of D-glucose: the open-chain form (A) and cyclic forms: α-D-glucopiranose (B) oraz β-D-glucopiranose (C)

In the solution, glucose exists mostly as a hemiacetal, *i.e.* in the cyclic form (Fig. 7.8.1b, 1c). The hemiacetal is formed by reaction of the aldehyde group of the C_1 carbon atom with the hydroxyl group of the C_5 carbon atom. Two possible products of such reaction for D-glucose are α- i β-D-glucopiranose, that exist in the equilibrium with the open-chain form in the aqueous solution.

Diabetes (*diabetes mellitus*) is a group of metabolic diseases in which the glucose excretion by the liver is elevated while its uptake by the other organs is decreased. There are two basic types of diabetes, *i.e.* type 1 diabetes, caused by degeneration of pancreas cells, that excrete insuline and type 2 diabetes. Type 2 diabetes, of unclear etiology, is chcracterized by hypeglycemia related to insuline resistance and relative lack of insullin. In the first phase of the pathology the increase of insuline level is observed followed by decrease of insuline exctreation due to significantly overloaded islet cells in the pancreas. The basic therapy for diabetes is insulino-therapy, based on administration of insuline or drugs elevating its level. Inproper dosage of these drugs as well as alcohol consumption or severe physical excercise may cause hypoglycaemia, *i.e.* the state of the organism in which the glucose blood level decreases below 55 mg dL^{-1} (3.0 mmol L^{-1}). In case of light hypoglycaemia this state can be tackled by ingestion or administration of glucose (5-20 gr) or carbohydrate foods. For severe hypoglycaemia, 20% glucose solution is intravenously administered followed by 10% glucose solution administered orally. Additionally, glucose is an adjuvant in numerous pharmaceuticals, such as Gastrolit, Litorsal, Orgalit, Glucardiamid.

AIM OF THE EXPERIMENT

The aim of the experiment is detection and determination of glucose in pharmaceuticals or body fluids and learning quantitative aspects of Raman measurements.

SCIENTIFIC BACKGROUND

1. Theoretical basis of Raman scattering; Chapter 2.
2. Building blocks of a dispersive and Fourier Raman spectrometer; Chapter 2.
3. Raman spectroscopy as a quantitative method.
4. Factors influencing Raman bands intensity.
5. *Diabetes mellitus.*

FURTHER READING

1. F.J. Holler, A.D. Skoog, S.R. Crouch: *Principles of experimental analysis*, Brooks/Cole Cengage Learning, Belmont, USA, 2007, pp. 481-493.

EQUIPMENT, MATERIALS, CHEMICALS

1. FT-Raman spectrometer, for instance Bruker MultiRAM and necessary software, for instance OPUS Bruker.
2. Laboratory dishes, wash bottles, automatic pipettes, measurement cuvettes, analytical balance.
3. Pure glucose.

4. Pharmaceuticals containing glucose, for instance Gastrolit, Litorsal, Orgalit, Glucardiamid or body fluids.

PROCEDURE

The experiment is based on measurement of Raman spectra of several aqueous solutions of glucose of known concentrations following measurement of Raman spectra of a body fluid(s) or an aqueous solution(s) of a medicine.

I. Preparation of solutions and spectra recording
   1. Place solid glucose in the measurement cell.
   2. Set up proper measurement parameters such as laser power, spectral resolution, and scan number.
   3. Focus laser on the sample, observe an interferogram for glucose.
   4. Record a Raman spectrum of solid glucose.
   5. Using an analytical balance weigh *ca.* 1.2 gr of glucose.
   6. Transfer the weighted glucose to a volumetric flask and dissolve in 4 ml of distilled water.
   7. Transfer 2 ml of the latter solution to a clean volumetric flask and dissolve with distilled water in the 2:1 ratio.
   8. Repeat point 7 threefold in order to obtain five aqueous solutions of glucose.
   9. Change the inlet for the one adjusted to measurements of liquids.
   10. Wash a measurement cuvette twofold using a small volume of a glucose solution.
   11. Transfer 500 µl of the latter solution to the cuvette.
   12. Place the measurement cuvette in the spectrometer chamber.
   13. Set up proper measurement parameters such as laser power, spectral resolution, and scan number and focus the laser on a sample.
   14. Record a Raman spectrum of a glucose solution.
   15. Repeat points 10–14 for all glucose solutions.
   16. Wash a measurement cuvette threefold with distilled water and then threefold using a small volume of an analyzed solution (a medicine solution or a body fluid).
      For a solid pharmaceutical, dissolve it in a small, known volume of distilled water.
   17. Transfer 500 µl of the analyzed solution to the cuvette.
   18. Record a Raman spectrum of the analyzed solution.
      Please note that the band intensity is equal to the integral of the area of the band.

II. Data analysis
   In the second part of the experiment, the analysis of the obtained experimental data will be performed. The set of Raman spectra of calibration solutions and the studied sample was obtained as the result of the measurement. The integral intensity of a marker glucose band will be calculated and plotted against the concentration of the solutions. The constructed calibration curve will be applied to calculate the concentration of glucose in the unknown solution(s).

1. Find a proper marker band for glucose and integrate its area in all obtained spectra by application of the same integration method and intervals.
2. Plot the function $I_{int}(c)$ and calculate the parameters of the line $I_{int} = ac + b$ using the linear regression method.
3. Calculate errors of a and b parameters and the correlation coefficient.
4. Based on the obtained relation $I_{int} = ac + b$ calculate the concentration of glucose in the studied solution(s).

REPORT

1. Based on recorded spectra note the Raman shifts of glucose characteristic bands.
2. Based on literature data, assign the most pronounced bands of glucose to the proper vibrations.
3. Present these data in the form of table 1:

Table 1. Wavenumbers of characteristic bands of glucose in aqueous solution with the assignment.

Wavenumber (cm⁻¹)	Assignment

4. Write down the calculations necessary to determine the glucose concentration in the studied sample(s):
   a. Present the data in the form of table 2:

Table 2. Integral intensity of a glucose marker band as a function of concentration of a solution.

Concentration (mol L⁻¹)	Integral Intensity (rel. units)

   b. Plot the graph $I_{int}(c)$.
   c. Write down the parameters of the equation $I_{int} = ac + b$, a and b errors and the correlation coefficient.
   d. Write down the obtained value of concentration of glucose in unknown solution(s).
5. Compare the obtained value(s) of glucose concentration with the reference value(s), given by a teacher.
6. Calculate the relative error of the measured value(s).
7. Based on obtained results, discuss the usefulness of Raman spectroscopy to determine the concentration of the analyte by considering both advantages and disadvantages of this method.

7.9. Resonance Raman scattering spectroscopy in hemoglobin structure studies

7.9. Resonance Raman scattering spectroscopy in hemoglobin structure studies

Jakub Dybaś, Antonina Chmura-Skirlińska, Katarzyna M. Marzec

7.9.1. Structure and physiology of hemoglobin

Hemoglobin is a protein which belongs to the group of porphyrins (hemoproteins). It occurrs in red blood cells and consists of protein part – the globin, and prosthetic group – the heme [1]. Hemoglobin is responsible for transporting of respiratory gases and takes part in maintaining of normal acid-base homeostasis.

The globin is tetrameric protein, containing two pair of subunits. The polypeptide chain of all subunits has 8 α-helices and is folded in seven places, what is a reason of a globular shape of such molecule [2,3]. In human hemoglobin there are a few different types of these subunits: α, β, γ, δ and ε. In adults there are almost exclusively hemoglobin A, consisting of two α and two β subunits (structure $\alpha_2\beta_2$) and small amount of hemoglobin A_2 of $\alpha_2\delta_2$ structure [1,2]. Each pair of the globin subunits, α and β, can be consider as a dimer, which is stabilized by a strong hydrophobic interactions. On the other hand weaker ionic interactions and hydrogen bonds exist between αβ dimers [4].

One molecule of heme is bounded to each of the subunits and composed of protoporphyrin and iron ion [1-4]. The protoporphyrin is composed of four pyrrole molecules bound together by methine bridges in tetrapyrrole ring, creating conjugated double bonds, what gives the heme intensive red colour. If iron ion is bounded to such protoporphiryn, it is called ferroprotoporphyrin. The most common is heme B, called also ferroprotoporphyrin IX, occurring, for example, in hemoglobin (Hb), myoglobin (Mb), peroxidases, catalases and cytochromes. Four methyl, two propanoic and two vinyl groups are bounded to the C_β carbon atoms of the pyrrole rings, see Fig. 7.9.1 [1]. The iron ion have six coordination sites. Four of them are occupied by nitrogens of the porphyrin ring (Fe-N), fifth take up the proximal histidine, His F8, (Fe-N_{His}), and sixth remains free [1,5]. In the neighbourhood of the last coordination site there is the distal histidine, His E7, which decrease affinity to carbon oxide and protect ferrous ion from the interaction with other hemoglobin molecules. Such contact could lead to oxidation of the iron ion to ferric form which is unable to bind oxygen [1,3,6] (Figures 7.9.1 and 7.9.2).

The iron ion inside the heme structure has mostly 23 (Fe^{3+}) or 24 (Fe^{2+}) electrons, from which 5 or 6 respectively, are valence electrons, located on *3d* orbitals (three bonding t_{2g} and two antibonding e_g^*). Depending on the manner of their distribution, *low spin* (LS) state and *high spin* (HS) state of the iron ion can be distinguished. In the LS state, electrons are coupled antiparallelly and located in orbitals of lower energy, and spin number for Fe^{2+} in this state is S = 0, while for Fe^{3+} S = 1/2. In the HS state of the iron ion electrons are coupled parallelly and for Fe^{2+} we have S = 2, and for Fe^{3+} S = 5/2 (Figure 7.9.3).

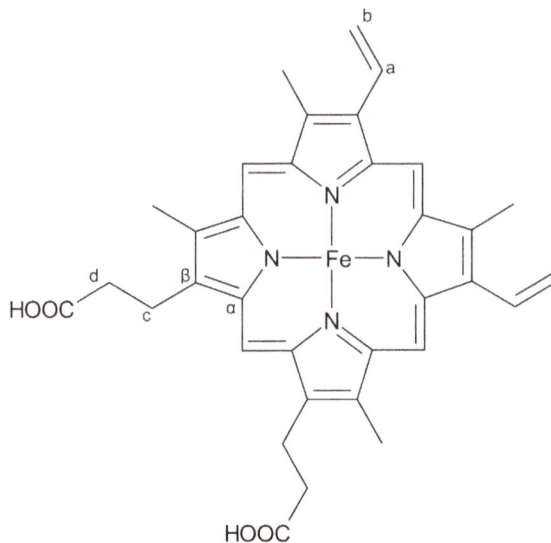

Fig. 7.9.1. Structure of heme; letter symbol are used in description of local coordinates in RRS: C_α, C_β – carbon atoms of the priol rings; $C_{\alpha'}$, C_β – carbon atoms of the methyl groups; C_c, C_d – carbon atoms of the propanoic groups [6].

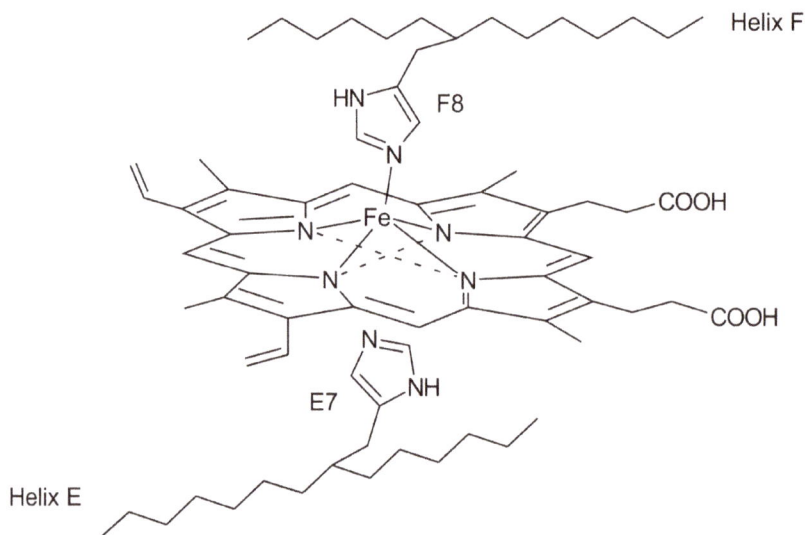

Fig. 7.9.2. Scheme of the heme plane (five-coordinated form of heme – typical for deoxyhemo-globin) and location of proximal (F8) and distal histidine (E7) [6]

The oxidation and the spin state is relevant to size of the iron ion. The Fe^{2+} is bigger from the Fe^{3+}, which has one electron less on the valence shell. In the LS state electron cloud is concentrated on three t_{2g} orbitals, while for the HS state electron cloud is spread. This has an impact on bigger size of the iron ion in HS state in comparison to LS state [7-9].

a)   b)

$S = 1/2$ $S = 0$

e_g^* $d_{x^2-y^2}$ d_{z^2}

t_{2g} d_{xy} d_{xz} d_{yz}

$S = 5/2$ $S = 2$

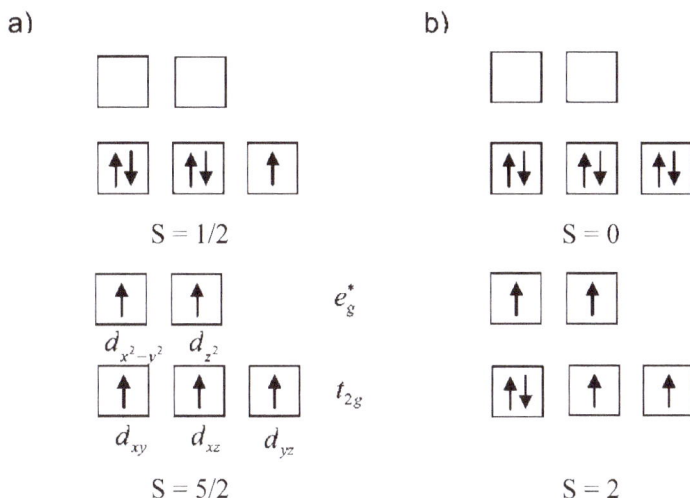

Fig. 7.9.3. Composition of electrons on 3d orbitals, a) LS and HS state of Fe^{3+}, b) LS and HS state of Fe^{2+}; S – spin quantum number [6]

The size of the iron ion have also significant influence on his position toward plane of porphyrin ring and size of whole Hb molecule. The smaller $_{LS}Fe^{3+}$ can „fit" in the crevice of the tetrapyrrole ring, and in fact remain in its plane, whereas bigger $_{HS}Fe^{3+}$ has to be out of the protoporphyrin plane, similarly to Fe^{2+} in both spin states [9,10].

Oxygen can bind only to these molecules of hemoglobin, in which the iron ion is on +2 oxidation state. Many sources claims, that binding of oxygen by hemoglobin should be called oxygenation, due to no iron ion oxidation state shift [1-3,11]. Actually, the situation is a little bit more complicated. During oxygen binding the displacement of electron density from Fe^{2+} toward the heme take place, and the partial negative charge appears at oxygen atom (superoxide anion is created). Therefore, the bond between the iron ion and oxygen should be consider as $Fe^{III}-O_2^-$ [6,12].

The oxygen free form of hemoglobin is pentacoordinated and is called deoxyhemoglobin, deoxyHb. The iron ion in deoxyHb is at HS and +2 oxidation state. The hemoglobin with oxygen is called oxyhemoglobin, oxyHb, and the iron ion in the centre of this form is at LS and +3 oxidation state. There is also the third form of Hb, unable to bind oxygen, in centre of which there is the HS iron ion at +3 oxidation state, and is called methemoglobin, metHb [6,12].

It is also worth to mention, that during binding of oxygen, the conformational shifts occur during which some part of hydrogen bonds and ionic interactions between αβ are ruptured. The created structure is characterized by bigger freedom of movement of the polypeptide chains, so it is called R form (*relaxed*), whereas in deoxyHb the subunits are more constrained, hence it is called T form (*taut*) [1,4].

7.9.2. Resonance Raman scattering spectroscopy in hemoglobin studies

All of the Raman techniques suits very well to biological system studies, because of no water scattering, what means no interruption in spectra interpretation, like in case of infrared spectroscopy. The resonance Raman spectroscopy (RRS) is commonly used in biomolecules studies due to much greater band intensities than in normal Raman spectroscopy (NR). Moreover it allows for a selective examination of different parts of molecular systems in dependence on used excitation wavelength, as described in Chapter 4.1.

From the RRS spectra not only information about structure of porphyrins and way of it's connection with protein can be obtained, but also information about bonded ligands, spin and oxidation state of metal ion. The frequencies of porphyrin systems vibrations are observe mostly in the range from 200 to 1700 cm^{-1}, while oscillations related to metal ion between 200 and 600 cm^{-1}[5]. The marker bands, allowing to distinguish the spin and the oxidation state, are mainly v_3, v_4, v_{10} and v_{19} bands [5-7]. The v_4 band at 1375 cm^{-1} (for Fe^{3+}) is connected with stretching vibration of the whole porphyrin ring, and reflects the oxidation state – it is shifted several cm^{-1} toward lower wavenumbers for Fe^{2+} (to about 1358 cm^{-1}) [5-7]. The shift in band position is caused by the effect of pulling out of π electrons from the porphyrins by metal-ligand bond. As a consequence the overlapping of d orbitals of the iron ion with π^* orbitals of porphyrin is observed. The v_3, v_{10} and v_{19} bands located above 1450 cm^{-1} are sensitive to deformation of the porphyrin ring, thus reflect the spin state of the iron ion [5-7]. Table 1 presents the assignment of the most

Table 7.9.1. Assignment of the most important RR bands and frequencies of vibrations on hemoglobin obtained with the use of 488, 514 and 633 nm laser wavelengths [6,13,14]

Assignment	Local coordinates[a]	Wavenumber [cm^{-1}]					
		OxyHb (488 nm)	DeoxyHb (488 nm)	OxyHb (514 nm)	DeoxyHb (514 nm)	OxyHb (633 nm)	DeoxyHb (633 nm)
	$v(Fe-O_2)$	572	–	572	–	–	–
v_7	$\delta(pyr\ def)_{sym}$	676	680	676	672	668	672
v_{15}	$v(pyr\ br)$	755	757	755	754	753	752
v_4	$v(pyr\ hr)_{sym}$	1376	1362	1371	1356	1367	1365
	$\delta(=CH_2)_{sym}$	1470	1474	1471	1471	–	–
$2v_{15}$	$v(pyr\ br)$	1505	–	–	–	–	–
v_2	$v(C_\beta C_\beta)$	1564	1560	–	–	1565	–
v_{37}	$v(C_\alpha C_m)_{as}$	1582	1584	1585	1580	–	1585
v_{19}	$v(C_\alpha C_m)_{as}$	1604	–	1604	1604	1604	1608
v_{10}	$v(C_\alpha C_m)_{as}$	1637	–	1638	–	1638	–

The mode notation is based on that proposed by Abe et al. [13] and Hu et al. [14].
[a] v – *streching*, δ – *scisoring*, def – *deformation*, br – *breathing*, hr – *half-ring*, as – *asymmetric*, sym – *symmetric*, pyr – *pyrrole*; symbols of carbon atoms are marked in the Fig. 7.9.1.

important bands for hemoglobin obtained with the use of 488, 514 and 633 nm laser wavelengths.

7.9.3. UV-Vis Absorption Spectrophotometry in hemoglobin studies

The absorption spectra of hemeproteins in visible range have mainly bands arise due to π-π^* electron shifts. The most abundant band is located at 410 nm and is called Soret band (or B band), from the name of its discoverer. At 500-600 nm region there is one or two bands, for Fe^{2+} and Fe^{3+} respectively, which are about ten times less intensive from the Soret band, and are called α (or Q_v) and β (or Q_0) bands. The first one arise due to vibronic coupling, and the second one also come from the π-π^* electron shifts [5]. In porhpyrin systems with Fe^{3+} at the HS there is observed additional band at 640 nm, related to charge transfer process (CT). The UV-Vis absorption spectrophotometry is thus fast and simple method to assess the oxidation and the spin state of the iron ion, confirming the data from the RRS spectra.

AIM OF THE EXPERIMENT

1. Recording spectra of the two forms of hemoglobin (oxyHb and deoxyHb) by using resonance Raman spectroscopy and UV-Vis absorption spectrophoto-metry.
2. Examination of used laser power (for one excitation wavelength) on the spectral profile of obtained Raman spectra of oxyHb.
3. An assignment of the marker bands for oxyHb and deoxyHb and comparison of bands positions of both Hb forms for two different excitation wavelengths. (In case of inability of using more than one excitation laser wavelength, Figure 7.9.5 presents comparison of oxyHb and deoxyHb Raman spectra, gained from 488nm excitation laser wavelength, which can be used in data analysis).

SCIENTIFIC BACKGROUND

1. Theory of the resonance Raman spectroscopy; Chapter 4.2.
2. Comparison of the NR with the RRS: advantages and disadvantages; Chapters 2 and 4.2.
3. Structure and physiology of hemoglobin.

EQUIPMENT, MATERIALS, CHEMICALS

1. Hemoglobin (solid, for microbiology) *i.e.* hemoglobin from bovine blood, FLUKA (CAS 9008-02-0); sodium dithionite (solid, tech. grade) *i.e.* from Sigma Aldrich (CAS 7775-14-6).
2. Raman spectrometer with the laser lines different than 488 nm (*i.e.* 514,5 and 632,8 nm or other from visible region).
3. UV-Vis spectrophotometer.
4. Software allowing to spectra creation and data analysis.

PROCEDURE

1. Prepare solutions:
   a. oxyHb – dissolve 6.5 mg of hemoglobin standard in 50 cm^3 of distilled water (2 μM/dm^3);
   b. deoxyHb – to 10 cm^3 solution of oxyHb add approximately 20 mg of sodium dithionate (prepare just before the measurement).
2. Record the UV-Vis absorption spectra of oxyHb and deyxHb solutions in the spectral region from 300 to 700 nm.
3. Record the RRS spectra with two excitation wavelength different from 488 nm (for example 514 nm and 633 nm or other from the visible range).
   a. By using one laser excitation wavelength record series of single spectra of hemoglobin standard crystals (oxyHb) using two different laser power (*i.e.* 10-100 μW and 1 mW). After measurements make average spectrum from gained single spectra for each of used laser power.
   b. By using laser excitation wavelength used in subparagraph b) and the lower laser power (about 10-100 μW):
      ■ record ten single Raman spectra of prepared oxyHb solution (number of scans as needed) – in this purpose, fill the cuvette with the oxyHb solution (Figure 7.9.4A), or place a drop of the solution on microscopic slide (Figure 7.9.4B);
      ■ record ten single Raman spectra of prepared deoxyHb solution as in subparagraph above.
   c. Repeat the measurements described in b) by using different excitation wavelength (or use spectra showed in Figure 7.9.5).
   d. Make the average Raman spectra of oxyHb and deoxyHb from a single Raman spectra obtained with the use of different laser excitation wavelength.

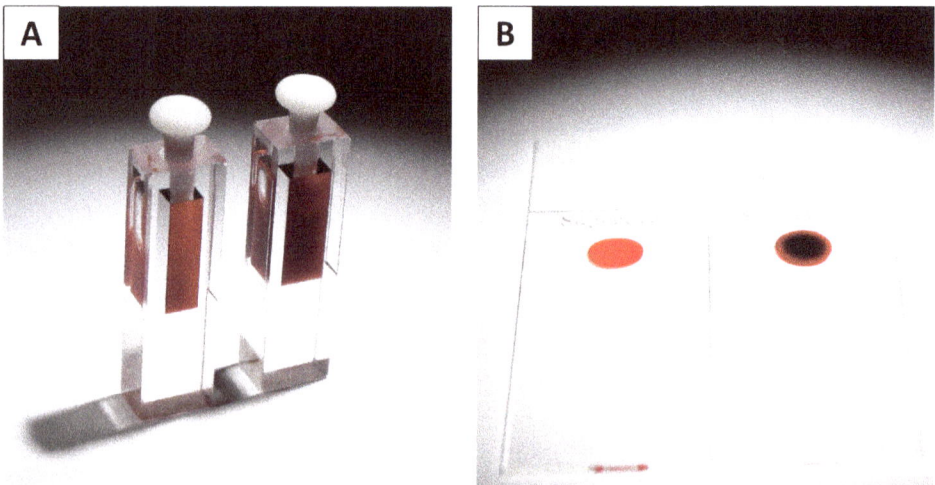

Fig. 7.9.4. The pictures of oxyHb and deoxyHb solution: a) in the cuvettes, b) on microscopic slide; oxyHb solution on the left, deoxyHb on the right

SUPPLEMENT

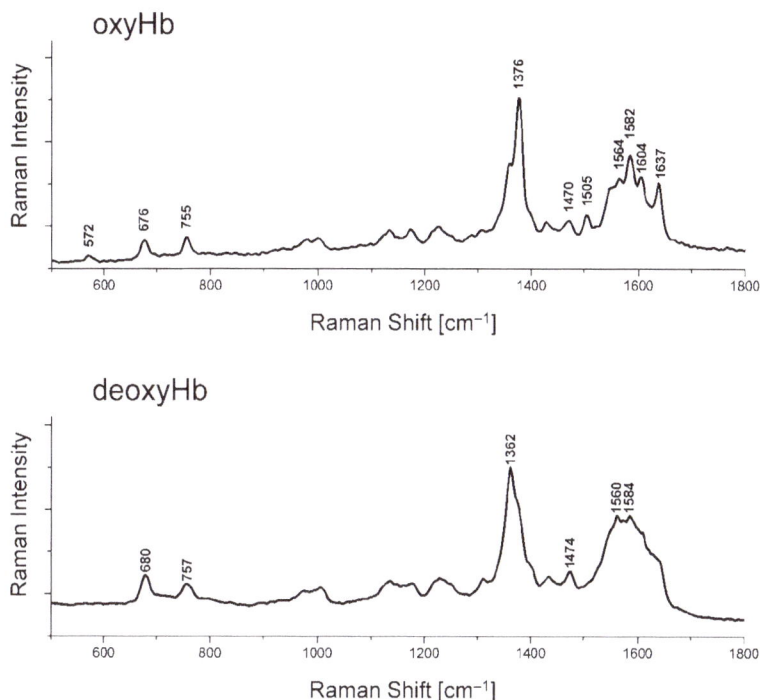

Fig 7.9.5. Resonance Raman Spectra of oxyHb and deoxyHb recorded by using 488 nm laser excitation wavelength

REPORT

1. Collect and describe recorded UV-Vis spectra of oxyHb and deoxyHb.
2. Present graphically and describe averaged RRS spectra:
   a. Compare Raman spectra recorded with different powers of the excitation source (as described in 2a) and explain the observed differences in spectral profile of oxyHb (photo/thermal laser induced photodissociation of O_2).
   b. Using Table 1 identify and assign marker bands for averaged Raman spectra of oxyHb and deoxyHb, recorded by using two excitation wavelength, and in correlation with UV-Vis spectrophotometry explain spectral and structural differences.
   c. Justify which laser excitation wavelength is better in distinction of oxyHb from deoxyHb.
6. Discuss and sum up carried experiments.

References

1. Strayer L., *Biochemistry*, New York: W.H. Freeman, 2002.
2. Kozik A., Turyna B., *Molekularne Podstawy Biologii*, Wydawnictwo ZK, Kraków 1996.

3. Hames B.D., Hooper N.M., *Biochemia Krótkie Wykłady*, Wydanie II, PWN, Warszawa 2007.
4. Champe P.C., Harvey R.A., Ferrier D.R., *Biochemistry Lippincott's Illustrated Reviews Series*, Third Edition, Lippincott Williams & Wilkins, 2005.
5. Twardowski J., Anzenbacher P., *Spektroskopia Ramana i podczerwieni w biologii*, PWN, Warszawa 1998.
6. Dybaś J., *Badania wpływu NO na stan funkcjonalny erytrocytów przy użyciu spektroskopii ramanowskiej*, MSc thesis, UJ, Krakow 2014 (in Polish).
7. Twardowski J., Proniewicz L.M., *Biospektroskopia Tom 4*, PWN, Warszawa 1990.
8. Crichton R., *Inorganic Biochemistry of Iron Metabolism*, Wiley, 2001.
9. Narayanan P., *Essentials of Biophysics*, New Age International, 2001.
10. Lewis D.F.V., *Cytochromes P-450*, Taylor & Francis, 1996.
11. Solomon E.P., Berg L.R., Martin D.W., *Biology*, Ninth Edition, Brooks/Cole, 2011.
12. Spiro T.G., Strekas T.C., *Resonance Raman spectra of heme proteins. effects of oxidation and spin state*, J. Am. Chem. Soc., **96**, 338 (1974).
13. Abe M., Kitagawa T., Kyogoku K., *Resonance Raman spectra of octaethyloporphyrinato-Ni(II) and meso-deuterated and ^{15}N substituted derivatives. II. A normal coordinate analysis*, J. Chem. Phys., **69**, 4526 (1978).
14. Hu S., Smith K.M., Spiro T.G., *Assignment of protoheme resonance Raman spectrum by heme labeling in myoglobin*, J. Am. Chem. Soc., **118**, 12638 (1996).

7.10. Identification of enantiomers of bornyl acetate and α-pinene in essential oils from the Siberian fir needles

Katarzyna Chruszcz-Lipska

7.10.1. Terpenes

The term "terpenes" originates from turpentine, a pleasant smelling viscous balsam distilled from resin, which itself is obtained by cutting or carving off the bark of numerous conifer tree species. Terpenes are natural compounds built up from isoprene (C_5H_8, 2-methyl-1,3-butadiene, Fig. 7.10.1) subunits. The basic molecular formula of terpenes is $(C_5H_8)_n$ where n is the number of linked isoprene residues (Table 7.10.1). This is the isoprene rule found by Ruzicka and Wallach. As a result terpenes are also denoted as isoprenoids [1].

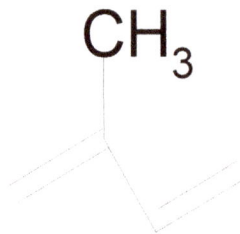

$$CH_3$$

Fig 7.10.1. Structure of isoprene

Table 7.10.1. Classification of terpenes by the number of isoprene units

Number of isoprene units (n)	Formula	Name	Example
1	C_5H_8	hemiterpenes	isoprene
2	$C_{10}H_{16}$	monoterpenes (terpenes)	α-pinene
3	$C_{15}H_{24}$	sesquiterpenes	farnesol
4	$C_{20}H_{32}$	diterpenes	phytol
5	$C_{25}H_{40}$	sesterterpenes	epiterpestacin
6	$C_{30}H_{48}$	triterpenes	squalene
8	$C_{40}H_{64}$	tetraterpenes	lycopene
n	$(C_5H_8)_n$	politerpenes	natural rubber

When terpenes are modified chemically and contain in additional groups in the structure such as -OH, -CHO, C=O, -COOH, -O-O- and/or takes place of the rearrangement of the carbon skeleton, the resulting compounds are referred to as terpenoids.

7.10.2. Stereochemistry of terpenes

In general, enantiomers (which are two stereoisomers that are mirror images of each other and are non-superposable), have identical chemical and physical properties with two exceptions *i.e.* -they cause rotation of the polarization plane of linear polarized light in opposite directions, to the left or to the right, with the same degree of angle, which is a characteristic feature of that particular substance, and, -they react with other optically active compounds in a different way.

A remarkable, example of such phenomenon of chirality and its effect are flavour compounds. Enantiomers of numerous natural compounds essential for aromas such as terpenes/oids have been described as having qualitatively or quantitatively different aromas. For example S-(+)-carvone smells like caraway, while its mirror image, R-(-)-carvone smells like spearmint (Fig. 2), S-(-)-limonene smells like turpentine, whereas R-(+)-limonene smells like orange [2]. Sometimes differences concern only to the aroma intensity like in the cause of α-pinene or camphor. There are also known examples where a dextrorotatory form of the compound has a distinct aroma, while its levorotary form is odourless or vice versa. These smells are different to the people because human olfactory receptors, which are chiral behave differently in the presence of different enantiomers. Additionally, research indicates that the different sensory properties of optical isomers of terpenes are connected with their structure. Strongly polar and bipolar compounds show significant differences in their sensory properties. For slightly polar or non-polar compounds, the differences in sensory properties between enantiomers are small or are not observed [3].

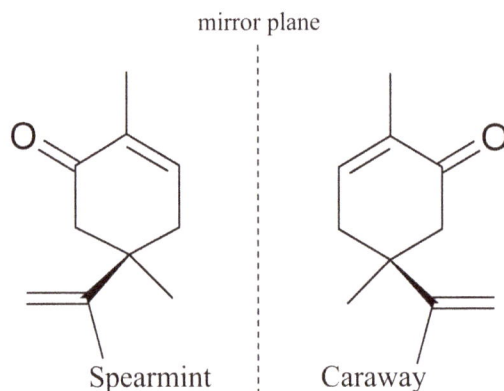

Fig. 7.10.2. Aroma of enantiomers of carvone

7.10.3. Enantiomeric contents of terpenes in essential oils

In general, terpenes are present in essential oils both in pure enantiomeric form and as a non-racemic mixture of two enantiomers. Measurement of enantiomeric occurrence of selected compounds in essential oils is an important factor for the characterization of a plant. Chiral analysis is applied to determine the authenticity of the essential oil. For example (-)-S-β-citronellol and (-)-(2S,4R)-cis and (-)-(2R,4R)-trans rose oxides are indicators of genuine rose oil, which is one of the most demanded and expensive oils in the world. The presence of racemic β-citronellol and rose oxides confirm the adulteration of the oil. In the same way, it can also be used to validate the authenticity of natural bergamot oil, which should contain only R-isomers of linalool and linalyl acetate. In addition, other essential oils have specific contents of selected optical active isomers [2].

From main conifer species, most essential oils which are produced today originate from the genus fir, pine, spruce, juniper and cedar (*Abies, Pinus, Picea, Juniperus* and *Cedrus*). What is interesting is the fact that the enantiomeric composition of terpenic hydrocarbons in essential oil of conifers depends not only on the species of the plant but also depends on the geographical region where that plant is growing. Currently, the most popular essential oil among the conifers is pichtae essential oil (Siberian fir needle oil, *Abies Sibirica* oil) which is primarily known for its anti-inflammatory and antispasmodic properties. Pichtae oil is very rich in bornyl acetate and contains 25-42% of this component. According to the literature *Abies Sibirica* oil and other oils from the species of conifers include solely the (-)-bornyl acetate [4,5]. On the other hand, other terpenic components of that oil exist in the sample in two enantiomeric forms with different percentage content. In Table 7.10.2 the comparison of the percentage composition of the enantiomers of selected compounds in *Abies Siberica* oil of different origin is presented [6].

Table 7.10.2. Enantiomeric contents [%] of selected components of pichta oils originated from different countries [6]

Compound	Austria	Italy	Korea
(+)-α-pinene	16.5	6.8	6.2
(-)-α-pinene	7.6	17.7	18.2
(+)-limonene	3.0	2.5	1.2
(-)-limonene	9.6	39.0	16.4
(+)-camphene	1.5	1.0	2.6
(-)-camphene	51.4	21.7	52.2

7.10.4. Raman optical activity spectra of terpenes

Small chiral molecules such as terpenes(oids) are excellent subjects for Raman optical activity study because they exhibit very strong ROA signals, are stable under high-power laser radiation and are often neat liquid at room temperature.

Figure 7.10.3 presents Raman and ROA spectra of two enantiomers of fenchone which are included in many plant species like fennel, lavender or tansy. Raman spectra of (+)-fenchone and (-)-fenchone are identical but ROA spectra of these enantiomers are mirror images. This means that ROA spectroscopy is sensitive on the absolute configuration of the compound.

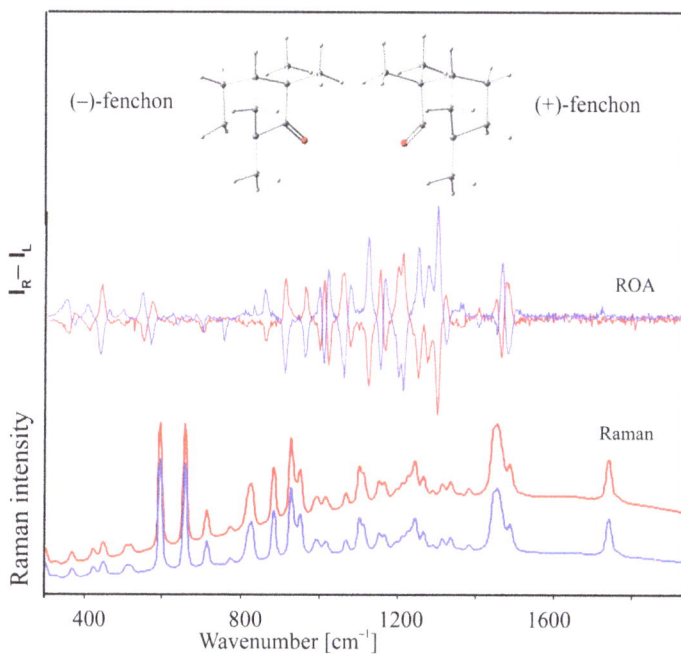

Fig. 7.10.3. Raman and ROA spectra of (+)-fenchone (red) and (-)-fenchone (blue)

It is widely known from the literature that the principal components of an essential oil can be well recognized by Raman spectroscopy. Fig. 7.10.4 demonstrates that also the ROA spectrum of essential oil exhibits characteristic key bands of its main constituents, *i.e.* anethole, estragole, fenchone and limonene [7]. However, information obtained from Raman and ROA spectra is a bit different, as the ROA spectrum leads information exclusively about optical active components of the essentials oil.

The signs of many observed bands (denoted as stars) in the ROA spectrum of commercially available fennel oil (Fig. 7.10.4) indicate the predominant or sole presence of (+)-fenchone in the sample. This data is fully consistent with the literature studies which indicate the presence of the pure dextrorotatory form of fenchone in fennel oil [8].

This example shows that ROA spectroscopy can be a powerful tool for *in situ* analysis of chiral components of essential oils, which are mixtures of numerous optical active compounds.

AIM OF THE EXPERIMENT

1. Collection of ROA spectra
2. Understanding of the nature of Raman and ROA spectra of enantiomers
3. Interpretation of ROA spectra of the mixture of optical active compounds.

SCIENTIFIC BACKGROUND

1. Fundamentals of Raman optical activity spectroscopy; Chapter 4.3.
2. Construction and operation of commercially available ROA spectrometer (SCP method); Chapter 4.3.
3. Basic information about terpenes.

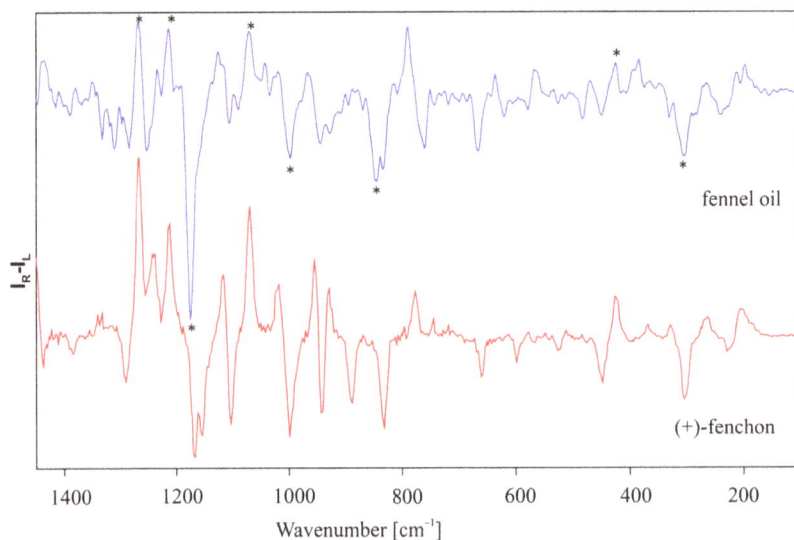

Fig. 7.10.4. Comparison of ROA spectra of fennel oil and (+)-fenchone

EQUIPMENT, MATERIALS, CHEMICALS

1. Chemicals: enantiomers of α-pinen and bornyl acetate, commercially available pichtae essential oil samples
2. Spectrometer of Raman optical activity.
3. Software for spectral data visualization (*i.e.* Excel, OPUS, OMNIC).

PROCEDURE

1. Measure the Raman and ROA spectra of the enantiomers of reference substances ((+) and (-)-α-pinen, (+) and (-)-bornyl acetate). Optimize measurement conditions (exposure time, laser power, number of scans).
2. Identify the marker bands for individual enantiomers of investigated terpenes.
3. Measure of Raman and ROA spectra of pichta essential oil sample.
4. Identify marker bands of individual enantiomers in the Raman and ROA spectra of essential oil sample.

REPORT

1. Indicate marker bands for enantiomers of investigated terpenes in the Raman and ROA reference spectra. On the basis of literature [5,9] try to assign the bands to the vibration of selected fragments of the molecules.
2. Based on the analysis of Raman and ROA spectra of pichta oil, discuss the occurrence of bornyl acetate and α-pinene enantiomers in pichta oil sample.
3. Discuss the utility of the ROA method to detect enantiomers in the essentials oil samples and other mixtures of optical active compounds.

References

1. Breitmaier E., Terpenes, WILEY-VCH Verlag GmbH & Co. KGaA, Weinheim, 2006.
2. Zawirska-Wojtasik R., *Chirality and the nature of food. Authenticity of aroma*, Acta Scientarum Polonorum, Technol. Aliment., **5**, 21 (2006).
3. Padrayuttawat A., Yoshizawa T., Tamura H., Tokunaga T., *Optical isomers and odor thresholds of volatile constituents* in: *Citrus sudachi, Food Science and Technology International, Tokyo*, **3**, 402 (1997).
4. Pureswaran D.S.,Gries R., Borden J.H., *Quantitative variation in monoterpenes in four species of conifers*, Biochem. System. Ecol., **32**, 1109 (2004).
5. Chruszcz-Lipska K., Blanch E., *In situ analysis of chiral components of pichtae essential oil by means of ROA spectroscopy: experimental and theoretical Raman and ROA spectra of bornyl acetate*, J. Raman Spectrosc., **43**, 286 (2012).
6. Ochocka J.R., Asztemborska M., Sybilska D., Langa W., *Determination of enantiomers of terpenie hydrocarbons In Essentials oils obtained from species of Pinus and Abies*, Pharmaceut. Biol., **40**, 395 (2002).
7. Aprotosoaie A.C., Spac A., Hancianu M., Miron A., Tanasescu V.F., Dorneanu V., Stanescu U., *The chemical profile of essential oils obtained from fennel fruits (Foeniculum vulgare mill.)*, Farmacia, **58**, 46 (2010).
8. Ravid U., Putievsky E., Katzir I., Ikan R., *Chiral GC analysis of enantiomerically pure fenchone in essential oils*, Flav. Fragrance J., **7**, 169 (2006).

9.   Bour P., Baumruk V., Hanzlikova J., *Measurement and calculation of the Raman optical acti-vity of a-pinene and trans-pinane*, Coll. Czech. Chem. Comm., **62**, 1384 (1997).

7.11. Determination of absolute configuration of α-pinene enantiomers by means of Raman optical activity and quantum chemical calculations

Grzegorz Zając, Małgorzata Barańska

α-Pinene is a cyclic monoterpene with a characteristic turpentine smell, it is found in the essential oils of coniferous trees, and used in cosmetics as well as in the pro-duction of different aroma compounds. At room temperature α-pinene is a liquid, which is great advantage in the ROA measurements, that needs high concentra-tions of the sample. In over a dozen minutes of the α-pinene measurements, it is possible to get a good quality ROA spectrum, with high signal to noise ratio. It makes α-pinene suitable compound for the use as a standard in verifying proper operation of the ROA spectrometer. ROA spectra of α-pinene were one of the first measured [1]. Figure 7.11.1 shows two enantiomers of α-pinene.

7.11.1. Theoretical calculations of ROA

Theoretical calculations of ROA are necessary in full analysis of experimental spec-tra. Comparison of the theoretical and experimental data in many cases allows for a clear determination of absolute configuration of chiral compounds. For molecules that have one pair of enantiomers, calculations only for one of them is needed, which is compared with the experimental spectrum. If the corresponding bands in experimental and theoretical spectra are matched (signs), it means that we measu-red and calculated the same enantiomer. If these spectra are mirror images of each other, it means that we used the structure of the second enantiomer in our calcu-lations [2].

Theoretical calculations of ROA intensities can be done by using Hartree-Fock method and Density Functional Theory (DFT). The second one is used more often. The most popular DFT functionals used in this case are hybrid ones, especially B3LYP and B3PW91. *Ab initio* calculations of ROA intensities are implemented in

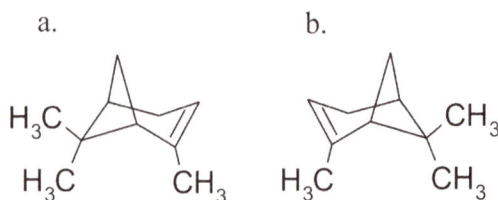

a.   b.

Fig. 7.11.1. Structural formulas of:
a) (+)-α-pinene, b) (–)-α-pinene

few programs, for example Gaussian [3]. Exemplary input file for Gaussian is presented below.

```
%chk=C:/name.chk              (location and name of checkpoint file)
%mem=8gb                      (memory)
%nproc=12                     (number of cores, processors)
#p B3LYP/6-31G(d) opt         (specification, method/basis set, calculation type)
(blank line)
Geometry optimization         (description)
(blank line)
0,1                           (charge, multiplicity)
 C  -2.30990362  -0.91083328  -2.20377359    (geometry)
(...)
(blank line)
```

In the calculations of ROA and Raman band intensities compared to the geometry optimization and the harmonic force field calculations, a different level of theory is needed. On the one hand the ROA calculations require additional diffuse functions, which definitely improves quality of obtained results, but on the other hand the complex basis sets like: 6-311++G(2d,2p) or aug-cc-pVTZ makes the calculations extremely time-consuming. So, a compromise between length and accuracy of the calculations is needed [4]. In 2004 the rDPS basis set appeared in the literature, which is based on 3-21++G basis set with additional p-type semi-diffuse function with 0.2 exponent on hydrogen atom. It gives comparable results to aug-cc-pVDZ and aug-cc-pVTZ, but reduction of computation time is remarkable [5].

Computation time could be reduced also in a two-step procedure. In the first step, for geometry optimization and harmonic force field, less complicated basis sets are used, for example cc-pVTZ or 6-31G(d). While in the second step, for ROA and Raman intensities, basis sets with diffuse augmentation can be applied, for example earlier discussed rDPS or aug(sp)-cc-pVDZ and aug-cc-pVDZ. Furthermore, in a one-step procedure, geometry optimization, harmonic force field and ROA intensities calculations have to be done on one theory level, which is suitable for ROA calculations. The one-step procedure is definitely more time-consuming, but quality of the obtained results is similar [4]. In Figure 7.11.2, (–)-α-pinene theoretical ROA spectra are shown, which are computed using B3LYP/aug-cc-pVDZ//B3LYP/aug-cc-pVDZ, B3LYP/aug-cc-pVDZ//B3LYP/6-31G(d), B3LYP/6-31G(d)//B3LYP/6-31G(d) levels of theory. Above notation will be used further. The first section shows the level of theory for ROA and Raman intensities, while the second one, (separated by //) the geometry optimization and harmonic force field calculations.

These calculations were done on the Zeus supercomputer in ACK Cyfronet AGH in Krakow, via PlGrid Infrastructure. Actual computation time was shorter, because 12 processors were used.

One can see that computation time and quality of spectra are changing using different theory levels. In the B3LYP/6-31G(d)//B3LYP/6-31G(d) spectrum, different

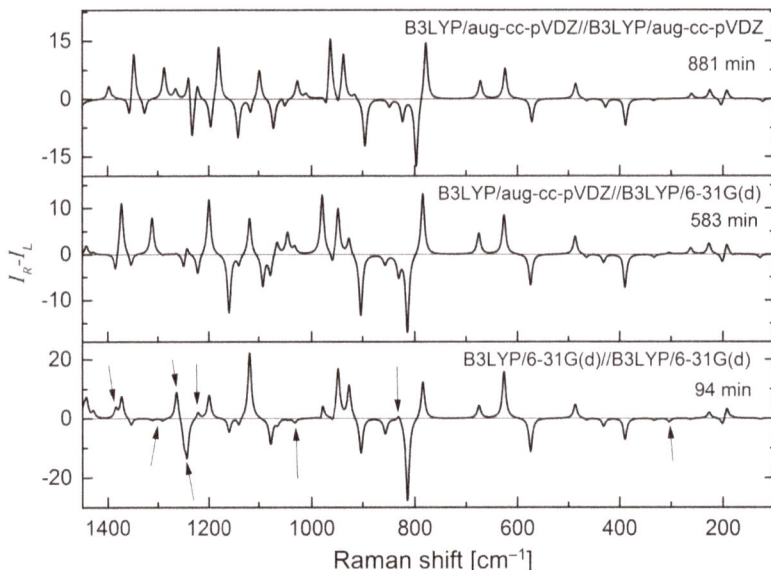

Fig. 7.11.2. Theoretical ROA spectra of (–)-α-pinene obtained by: B3LYP/aug-cc-pVDZ//B3LYP/aug-cc-pVDZ, B3LYP/aug-cc-pVDZ//B3LYP/6-31G(d), B3LYP/6-31G(d)//B3LYP/6-31G(d) levels of theory. Job CPU times from Gaussian output files are also given

signs of bands are matched (in respect to the spectra calculated at the higher theory levels), which confirms that basis sets without diffuse augmentation could give different/erroneous results. The B3LYP/aug-cc-pVDZ//B3LYP/aug-cc-pVDZ and B3LYP/aug-cc-pVDZ//B3LYP/6-31G(d) spectra are similar, so one and two-step procedure give similar results, but two-step procedure is definitely faster. A small shift of the bands maxima in these two spectra is caused by applying different levels of theory for harmonic force field calculations.

Input files for one and two-step procedure are shown below:
One-step procedure:

```
#p ,ethod/basis opt freq=roa cphf=rdfreq
```

(*description*)

```
0,1
```
(*geometry*)

```
532 nm
```
        (*incident light wavelength*)

Two-steps procedure
First step:

```
#p method/basis opt freq=noraman
```

(*description*)

```
0,1
```
(geometry)

Second step:

```
#p method/basis polar-roa cphf=rdfreq geom=check
```

(description)

```
0,1
```

```
532 nm
```
       (incident light wavelength)

opt   – geometry optimization.
polar=roa – ROA intensities calculations, harmonic force field constants are
             taken from checkpoint file.
cphf=rdfreq – dynamical, frequency dependent algorithm for solving CPHF
             equations (*Coupled Perturbed Hartree-Fock*). One line below the
             geometry, the incident light wavelength has to be written.
freq=roa   – harmonic force field and ROA intensities calculations.
freq=noraman – harmonic force field calculations without Raman activities.
geom=check – geometry will be taken from checkpoint file.

Detailed information about quantum chemical methods, basis sets, calculations of ROA intensities and Gaussian program technical support can be find in literature below [6-9].

AIM OF THE EXPERIMENT

The aim of this exercise is determination of absolute configuration of α-pinene enantiomers by using ROA measurements and quantum chemical calculations. The exercise is divided into two parts: experimental ROA measurements of (+) and (–)-α-pinene, and theoretical calculations of (–)-α-pinene, which are based on geometry optimization, harmonic force field calculations, and ROA and Raman intensities calculations (two-step procedure).

SCIENTIFIC BACKGROUND

1. Fundamentals of Raman Optical Activity spectroscopy; Chapter 4.3.

EQUIPMENT, MATERIALS, CHEMICALS

1. Spectrometer Chiral*RAMAN-2X*™ from BioTools, equipped with 532 nm laser (Nd:YAG second harmonic) and CCDsp™ camera.
2. Softwares Critical Link, LLC.
3. (+)-α-pinene, (–)-α-pinene (p.a.).
4. Computational software (e.g. Gaussian 09 D.01).
5. Software for visualization of calculations (e.g. GaussView 5).

PROCEDURE

I. Theoretical calculations
    1. Draw the molecular structure of (–)-α-pinene in the GaussView software.
    2. Write the input file for the first step calculations (geometry optimization and harmonic force field calculations) by using the B3LYP functional and 6-31G(d) basis set.
    3. Submit calculations.
    4. Write the input file for the second step calculations (ROA and Raman intensities) by using the B3LYP functional and rDPS basis set. Read geometry and harmonic force field constants from the previous first step calculations.
    5. Submit calculations, when first step is terminated normally.

The rDPS basis set is not implemented in the Gaussian 09 D.01. so we must enter it manually. In the input file, instead of basis set command, Gen keyword has to be written, which gives us possibility to define basis set at the end of the input file. On the https://bse.pnl.gov/bse/portal website you can find specification for many basis sets for selected atoms. In this case, you would need specification of 3-21++G basis set for hydrogen and carbon atom. Then you have to add p-type semi-diffuse function on hydrogen atom with 0.2 exponent. The input file with rDPS basis set is presented below.

```
#p B3LYP/Gen Polar=roa cphf=rdfreq geom=check
```

(description)

```
0,1

532 nm

H       0
S   2      1.00
           5.4471780     0.1562850
           0.8245470     0.9046910
S   1      1.00
           0.1831920     1.0000000
S   1      1.00
           0.0360000     1.0000000
P   1      1.00
           0.2000000     1.0000000
****
C       0
S   3      1.00
         172.2560000     0.0617669
          25.9109000     0.3587940
```

```
          5.5333500       0.7007130
SP  2     1.00
          3.6649800      -0.3958970       0.2364600
          0.7705450       1.2158400       0.8606190
SP  1     1.00
          0.1958570       1.0000000       1.0000000
SP  1     1.00
          0.0438000       1.0000000       1.0000000
****
```

II. ROA measurement

1. Pipette 0.1 ml of (+)-α-pinene into the quartz microcuvette, clean cuvette using the optical leans cleaning tissue and methanol, and place in the sample holder in ROA spectrometer.
2. Chose measurement conditions, for α-pinene and Chiral*RAMAN-2X* spectrometer the optimal conditions are: 0.51 s of length of illumination period and 60 mW power of laser. These conditions may change, and it depends on the laser, spectrometer, cuvette, and purity of compound.
3. When measurement conditions are chosen, press "Accumulate ROA". 1024 scans are enough to get good quality ROA spectrum of α-pinene.
4. Repeat this procedure for (–)-α-pinene.

REPORT

1. Looking at the theoretical data, select the most important (the most intense) bands and assign molecule vibrational modes. Collect results in a table.
2. Set together experimental and theoretical spectra of α-pinene enantiomers, assign corresponding bands from theoretical and experimental spectra, compare signs of bands, and determine the absolute configuration.
3. Compare the obtained theoretical spectrum (vibration frequency and signs) and CPU time with the spectra demonstrated in the Figure 7.11.2. Discuss it.

References

1. Hug W., Kint S., Bailey G.F., Scherer J.R., *Raman Circular Intensity Differential Spectroscopy. Spectra of (-)-a-Pinene and (+)-a-Phenylethylamine*, JACS, **97**, 5589 (1975).
2. Nafie, L.A. *Vibrational Optical Activity: Principles and Applications*, John Wiley & Sons, Chichester, UK 2011.
3. Frisch, M.J., et al., *GAUSSIAN 09, Rev. D.01*, Gaussian Inc., Wallingford, CT, USA 2009.
4. Cheesman J.R., Frish M.J., *Basis Set Dependence of Vibrational Raman and Raman Optical Activity Intensities*, J. Chem. Theory Comput., **7**, 3323 (2011).
5. Zuber G., Hug W., *Rarefied Basis Sets for the Calculation of Optical Tensors. 1. The Importance of Gradients on Hydrogen Atoms for the Raman Scattering Tensor*, J. Phys. Chem. A, **108**, 2108 (2004).
6. Jensen F., *Introduction to Computational Chemistry, Second Edition*, John Wiley & Sons, Chichester, UK 2007.

7. Ruud K., Thorlvaldsen A.J., *Theoretical Approaches to the Calculation of Raman Optical Activity Spectra*, Chirality, **21**, E54 (2009).
8. Pecul M., Ruud K., *Ab Initio Calculation of Vibrational Raman Optical Activity*, Intern. J. Quant. Chem., **104**, 816 (2005).
9. http://www.gaussian.com/g_tech/1.htm.

7.12. Estimation of surface enhancement factor and adsorption studies of 3-amino-5-mercapto--1,2,4-triazole (AMT) on silver using surface-enhanced Raman scattering spectroscopy (SERS)

Agata Królikowska, Jolanta Bukowska

7.12.1. Surface enhancement factor in SERS

Estimate of magnitude of surface enhancement factor in *surface-enhanced Raman scattering* (SERS, *see* Chapter 4.2) is a difficult issue already since the discovery of the increased intensity of Raman scattering signal for the molecules adsorbed on metallic nanostructures (mostly Ag and Au), comparing to conventional Raman experiment [1]. Initial and unfortunately incorrect interpretation of the nature of SERS effect attributed enhancement of the signal to the increased number of the molecules on electrochemically roughened substrate (rich in the fractal-like structures). Afterwards it was demonstrated that stronger SERS signal is a consequence of the increased Raman scattering cross-section of studied molecules, which resulted in introduction of *enhancement factor* (EF) concept.

Enhancement factor is one the most important numbers characterizing quantitatively SERS effect. It allows preliminary estimate of SERS spectroscopy applicability to a given task. For example, it is typically assumed that single molecule detection with SERS is possible for EF in the range of 10^{14}. It means that in practice a purely electromagnetic enhancement is not sufficient, as its contribution in a total SERS enhancement factor is believed to be at most around 10^{10}. Therefore an additional enhancement is essential to enable detection of SERS signal from single molecules. It can arise from so called chemical effect (CT effect) or combining SERS and resonance Raman effect and thus spectrum excitation in the energy range corresponding to electronic absorption of adsorbed molecules (SERRS technique).

Magnitude of SERS enhancement factor is dependent on experimental conditions, such as:

- used SERS active substrates (*i.e.* type of metal, nanostructure geometry),
- excitation line (*i.e.* wavelength, polarization, angle of incidence),
- type of analyte (Raman scattering cross-section, mode of adsorption).

7.12.1.1. Types of definition of enhancement factor

Literature reports on the determined values of average EFs show wide discrepancy: typical values are in the range of 10^4-10^6, reaching even 10^8, but even papers claiming to obtain enhancement factors as high as 10^{14} can be found, not limited to the case of single molecule SERS [2]. Such a large diversity of enhancement factor values occurs mostly for two reasons:

- differently defined enhancement factors,
- practical method of EF estimation.

Variety of the conditions of SERS experiment *i.e.*: analysis of single or many molecule events, uncertainty of real surface concentration and its incorrect evaluation, averaging signal over time, spatial distribution and molecular orientation of the analyte make introduction of single EF definition practically impossible. Here three the most important definitions of SERS enhancement factors will be discussed.

7.12.1.2. Enhancement factor for a given substrate (*EF*)

From a practical point of view, defining SERS enhancement factor allowing comparison of average EF for various substrates is essential. The most common definition of an average SERS enhancement factor for a particular substrate can be given as (7.12.1) [2]

$$EF = \frac{I_{SERS}/N_{Surf}}{I_{RS}/N_{Vol}}, \qquad (7.12.1)$$

where I_{SERS} and I_{RS} denote respectively intensity of SERS signal and normal Raman scattering signal (for a given analyte), while N_{Surf} and N_{Vol} are respectively average molecule numbers in a scattering volume for SERS experiment conditions and for Raman scattering by a volumetric sample.

Substrate related SERS enhancement factor can be also expressed more rigorously in terms of experimentally measured signals (7.12.2) [2]:

$$EF = \frac{I_{SERS}/\mu_M \mu_S A_M}{I_{RS}/c_{RS} h_{eff}}. \qquad (7.12.2)$$

In formula (7.12.2), μ_M [m^{-2}] and μ_S [mol $\cdot$ m^{-2}] denote respectively surface density of individual SERS active nanostructures and molecules on the metal (for periodic structures), A_M [m^2] is a surface area of metallic nanostructure, c_{RS} [mol m^{-3}] corresponds to volumetric concentration of the sample used for conventional Raman scattering experiment, while h_{eff} [m] is an effective height of the scattering volume. Expression given by equation (7.12.2) can be reduced to this given by equation (7.12.1), when $N_{Surf} = \mu_M \mu_S A_M A_{eff}$ and $N_{Vol} = c_{RS} V = c_{RS} h_{eff} A_{eff}$, where A_{eff} is an effective area of a scattering volume (V [m^3]). EFs derived with a non-rigorous definition given by equation (7.12.1) may lead to artificial variations by as much as two orders of magnitude [2].

7.12.1.3. Analytical enhancement factor (AEF)

For many applications, evaluation of SERS signals amplification in comparison to normal Raman scattering (RS) under given experimental conditions is sufficient. Such approach is particularly useful in analytical chemistry.

Let us consider than analyte sample of concentration c_{RS} is giving normal Raman signal of intensity I_{RS}. Next, under the same conditions (energy and power of the laser, parameters of microscopic objective etc.) and for unchanged preparation conditions, the same analyte produces SERS signal of intensity I_{SERS}, for possibly different concentration given by c_{SERS}. The *analytical enhancement factor* (AEF) can be then defined as (7.12.3) [2]:

$$AEF = \frac{I_{SERS}/c_{SERS}}{I_{RS}/c_{RS}}. \tag{7.12.3}$$

Certainly, such defined enhancement factor is strongly dependent on the molecular adsorption mode, surface coverage (monolayer *vs* multilayer) and method of preparation of two-dimensional nanostructures. It does not allow comparison of different SERS substrate performances. Use of AEF is particularly relevant for SERS measurements in liquids, namely suspensions of colloidal metallic nanoparticles.

7.12.1.4. Single molecule enhancement factor (SMEF)

SERS signal for single molecules arises practically only from the molecules adsorbed at *hot spots*, which are the sites of metallic nanoparticles exhibiting exceptionally intense electromagnetic field. Rigorous definition of enhancement factor experienced by a given molecule at a specific point is difficult (called *single molecule enhancement factor*; SMEF), as it is determined by:

- polarizability tensor of the molecule,
- molecular orientation with respect to the surface and thus with respect to the orientation of the local electromagnetic field,
- orientation of the substrate with respect to the direction and polarization of the laser beam.

It is useful to define single molecule enhancement factor using ratio of the SERS intensity for a considered single molecule I_{SERS}^{SM} and average Raman scattering signal per molecule for the same probe, taken over all possible (random) orientations of the molecule in space $\langle I_{RS}^{SM} \rangle$ (7.12.4) [2].

$$SMEF = \frac{I_{SERS}^{SM}}{\langle I_{RS}^{SM} \rangle}. \tag{7.12.4}$$

Magnitude of $\langle I_{RS}^{SM} \rangle$ can be estimated experimentally from measurement of Raman scattering intensity for a solution of known concentration (for a given solvent).

7.12.2. The most common sources of errors in the determined values of SERS enhancement factors

In the literature many reports on surprisingly large SERS enhancement factors for novel SERS active nanostructures or in case of single molecule detection can be found. The main sources of these abnormalities are as follows [2]:

- ill-defined EF, including total ignoring the necessity of normalization with respect to normal Raman conditions,
- underestimation of the bare Raman cross-sections in case of measurements of dyes under resonance or pre-resonance conditions,
- difficulties with analysing SERS signal exclusively due to single molecules (except for the ultralow concentration approach).

For all these reasons some of the literature values of SERS EFs should be treated sceptically and a careful inspection of the method of their estimation and used experimental conditions should be performed.

7.12.3. Structure and properties of 3-amino-5-mercapto-1,2,4-triazole (AMT)

3-amino-5-mercapto-1,2,4-triazole (AMT, structural formula shown in Fig. 7.12.1) is a member of biologically important species, containing a 1,2,4-triazole moiety, possessing three nitrogen hetero-atoms in a five-membered ring. They are used extensively as drugs demonstrating antifungal, antibacterial, antidepressant and even anticancer properties [3].

Moreover, AMT as a heterocyclic compound containing thiol moiety (-SH) is capable of chemisorption on metals like Ag and Au (providing high SERS enhancement) and formation of densely packed monolayers. Such layers can be protective against corrosion and serve as a linker and electron transform promoter between the metal and biologically active redox species.

AMT molecule shows also a thiol-thione and amino-imine tautomerism. A deprotonation of both thiol group and ring nitrogen is also possible, which results again in a formation of various tautomeric forms. Described tautomers are shown in a scheme in Fig. 7.12.1. Presence of amino group, heterocyclic nitrogen atoms and variety of tautomeric forms make AMT adsorption on metal surface in a thiolate form not that obvious.

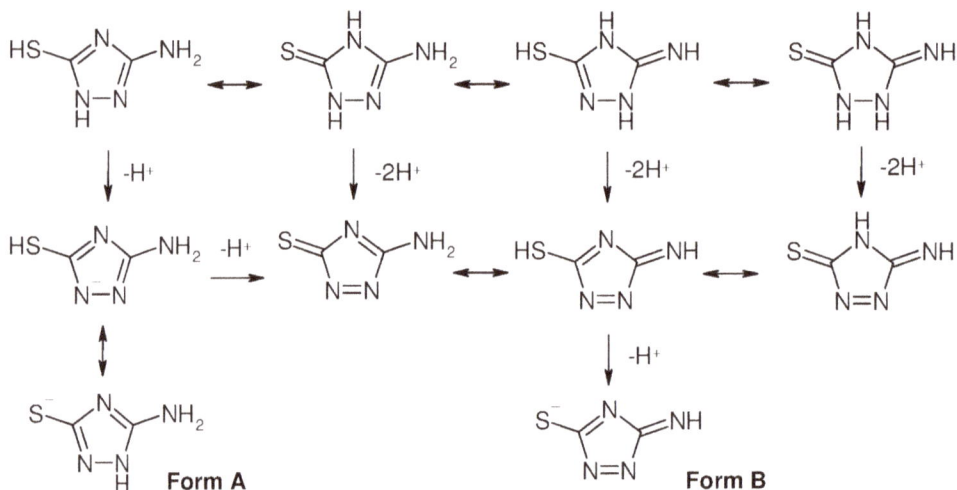

Fig. 7.12.1. Structural formula of 3-amino-5-mercapto-1,2,4-triazole (AMT) and thiol-thione and amino-imine tautomerism of AMT and its deprotonated forms [4] (form A and B explained further in the text)

7.12.4. SERS spectrum of 3-amino-5-mercapto-1,2,4-triazole (AMT) on silver substrate

SERS spectroscopy provides insight into monolayer structure, its chemical state, nature of interactions with the metal and molecular orientation with respect to the surface (with the use of surface selection rules), hence being and excellent tool for the evaluation of its potential toward a given application.

SERS studies for AMT adsorbed on silver, confirmed with XPS (*X-ray photoelectron spectroscopy*) results [4], showed that adsorption proceeds mostly with a thiolate formation, with a minimum contribution of a thione form. This conclusion was derived from the SERS enhancement of the stretching vibration of a C-S moiety around 480 cm^{-1} and the lack of the S-H stretching vibration around 2500 cm^{-1}. In SERS spectra two different AMT forms within a monolayer were identified [4], dependent on the experimental conditions: referred further as form A and B. Form A corresponds to AMT molecules with the non-deprotonated NH$_2$ groups, while form B was attributed to deprotonated form with imine H–N=C moiety (details of the structures are shown in Fig. 7.12.1). SERS spectra representative for A and B form are shown in Fig. 7.12.2. Non-deprotonated molecules (form A) form a monolayer titled with respect to the surface, stabilized by intermolecular hydrogen bonding (within a monolayer and with solution molecules) and π-electrons stacking interactions. Interactions of the ring with the metal are favoured for the deprotonated form (form B), leading to the parallel orientation of the molecules in the monolayer.

Transformation between A and B form is evidenced with striking changes in SERS spectrum of AMT on Ag: band around 1330 cm^{-1} correspond to A form, while SERS mode around 1360 cm^{-1} is a marker band of form B (see Fig. 7.12.2). Both these bands are due to the vibrational modes of a triazole ring [6]. Moreover, a relative intensity of v(C-S) band around 480 cm^{-1} is increased for A form, which using surface selection rules pointed at more upright molecular orientation of AMT molecules [4]. An applicability of surface selection rules to AMT on Ag is however limited, as there is a strong contribution of *charge transfer* effect (enhancement by chemical effect), as SERS spectrum is detectable even for smooth substrates.

A relative amount of A and B form in AMT monolayer on Ag is dependent on AMT concentration in solution, pH of surroundings and applied potential. Form A is stabilized by high AMT concentration, low pH values and lowering the potential [4].

AIM OF THE EXPERIMENT
The aim of the exercise is to gain abilities of:

1. Correct estimation of SERS enhancement factor.

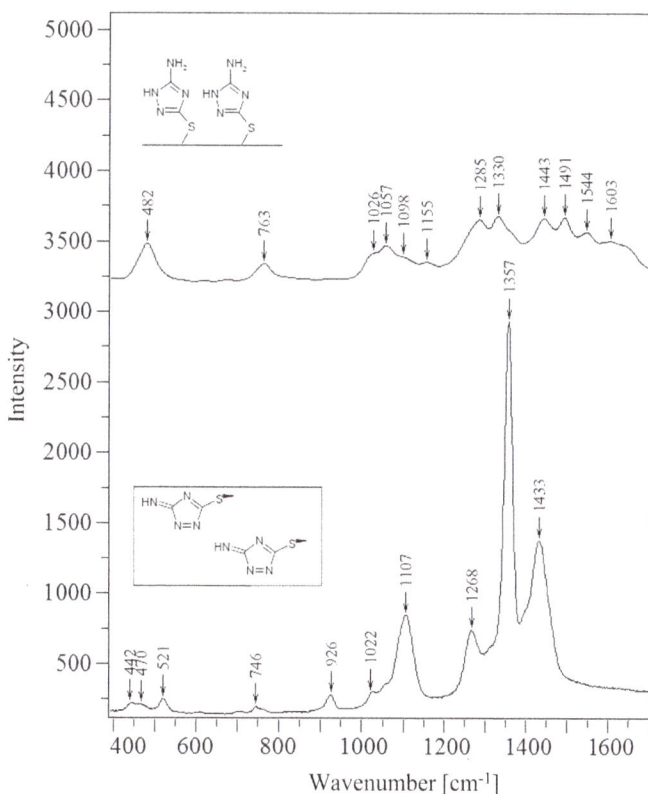

Fig. 7.12.2. SERS spectra (excitation 532nm) for two different adsorption modes of AMT on Ag surface: form A (top spectrum) and form B (bottom spectrum). Inset: structure of form A (non-deprotonated, upright orientation) and form B (deprotonated, lying flat) [5]

2. Preparation of SERS active substrate by electrochemical roughening of silver substrate.
3. Methodology of normal Raman scattering (RS) and SERS (on a roughened Ag plate) spectra collection.
4. Interpretation of SERS spectra for thin organic films, including an attempt of evaluation of adsorption mode and comparison with RS spectrum for volumetric sample.
5. Analysis of effect of experimental conditions (like concentration of the sample, pH of the surroundings) on structure of the adsorbate, by examining SERS signal.

SCIENTIFIC BACKGROUND

1. Raman scattering – physical fundamentals and selection rules; Chapter 2
2. Differences between normal Raman scattering and SERS spectrum; Chapter 4.2.
3. Theories explaining mechanism of surface enhancement in SERS: enhancement by electromagnetic field and chemical effect; Chapter 4.2.
4. Types of SERS active substrates.
5. Factors affecting SERS spectrum.
6. Definitions of surface enhancement factor in SERS: EF, AEF and SMEF.

EQUIPMENT, MATERIALS, CHEMICALS

1. Chemicals: amino-5-mercapto-1,2,4-triazole, NaOH, KCl, acetone, phosphate buffers (AR; analytical grade).
2. Silver plate, reference electrode and counter electrode (Pt sheet).
3. Potentiostat – galvanostat.
4. Raman spectrometer, coupled with a confocal microscope.
5. Software for data treatment.

PROCEDURE

1. Roughening of Ag plates. Degreasing three Ag plates with acetone. Annealing Pt sheet in a burner flame. Mounting a three-electrode electrochemical system, with Ag as working electrode, Pt as a counter electrode and a silver/silver chloride as reference electrode. Connecting electrodes with a potentiostat-galvanostat and selection of cyclic voltammetry technique. Five oxidation-reduction cycles in which the potential is changed from +300mV to -300 mV (*vs* Ag/AgCl 1M KCl$_{aq}$), at a sweep rate of 5 mV/s should be applied to roughen silver and a potential held for 30s in a negative vortex for the terminal cycle.
2. Adsorption of AMT on one roughened silver plate by immersion in 0.3 M AMT solution in 0.4 M NaOH for about 1 hour.
3. Collection of RS for volumetric sample of 0.3 M AMT solution in 0.4 M NaOH.
4. Collection of *in situ* and *ex situ* SERS spectra for AMT monolayer grown on Ag surface from alkaline solution.
5. Adsorption of AMT on two remaining Ag plates by immersion in 0.01 M aqueous AMT solutions of pH about 2.2 and 12.7 (in phosphate buffers) for at least 1 hour.

6. Collection of *in situ* SERS spectra for AMT monolayers formed under different pH.
7. Selection and assignment of diagnostic bands to AMT molecular vibrations, discussed in section 7.12.4.

REPORT

1. Describe and interpret differences between RS spectrum of AMT in alkaline solution and SERS spectrum of the monolayer adsorbed from analogous solution.
2. Select AMT diagnostic bands of for the spectra mentioned above (SERS counterparts of RS bands) and measure their intensities in order to estimate enhancement factors EF given by equation (7.12.1).
   Caution: Estimation of the number of the molecules in scattering volume for the volumetric sample and adsorbate should be performed on the base of AMT concentration is solution, surface concentration Γ (for AMT monolayer $\Gamma = 0.5$ nmol cm^{-2} can be assumed) and parameters of used confocal Raman microscope (h_{eff} of the applied objective):
   $N_{Surf} = \Gamma\, A_{eff}$, while $N_{Vol} = c_{RS}\, h_{eff}\, A_{eff}$.
3. Assign the most important SERS bands of AMT to the vibrational modes of the molecule [4,6]. Evaluate a chemical state for AMT molecules bound to silver surface by analysis of SERS spectra for studied samples [4].
4. Examine effect of pH, AMT concentration and presence of the monomer on orientation of the adsorbate molecules (parallel, upright or intermediate) with respect to the metal surface, based on the analysis of collected SERS spectra and compare with literature results [4].
5. Make a report including aim of the exercise, experimental procedure, discussion of the results and conclusions.

References

1. Fleischmann M., Hendra P.J., McQuillan A.J., *Raman spectra of pyridine at a silver electrode*, Chem. Phys. Lett., **26**, 163 (1974).
2. Le Ru E. C., Blackie E., Meyer M., Etchegoin P. G., *Surface Enhanced Raman Scattering Enhancement Factors: A Comprehensive Study*, J. Phys. Chem. C, **111**, 13794 (2007).
3. Singh R., Chouchan A., Important methods of synthesis and biological significance of 1,2,4-triazoles, World Journal of Pharmacy and Pharmaceutical Science, **3**, 874 (2014).
4. Wrzosek B., Bukowska J., *Molecular Structure of 3-Amino-5-mercapto-1,2,4-triazole Self--Assembled Monolayers on Ag and Au Surfaces*, J. Phys. Chem. C, **111**, 7397 (2007).
5. Piotrowski P., Wrzosek B, Królikowska A., Bukowska J., *A SERS-based pH sensor utilizing 3-amino-5-mercapto-1, 2, 4-triazole functionalized Ag nanoparticles*, Analyst, **139**, 1101 (2014).
6. Wrzosek B., Cukras J., Bukowska J., *Adsorption of 1,2,4-triazole on a silver electrode: surface-enhanced Raman spectroscopy and density functional theory studies*, J. Raman Spectrosc., **43**, 1010 (2012).

7.13. Identification and studying of distribution of caffeine in drug samples *in situ*

Małgorzata Barańska, Agnieszka Kaczor, Kamilla Malek

Bioactive compounds can be studied in their natural environment, in the isolated form or in the processed form in pharmaceuticals. Raman imaging has a growing potential in the pharmaceutical applications and are increasingly used in pharmacy, among others to monitor technological production process and quality control of the medicines. The reason of this growing popularity is the fact that Raman imaging enables obtaining information about molecular structure and distribution of an active substance in the medicine at the same time.

Particularly, this method enables *in situ* identification of the active substance (for instance in a tablet) occurring together with the fillers and flavouring compounds. One can obtain the knowledge about the analyte's distribution, verify the distribution homogeneity and determine the size of the active substance's grains. The latter parameter is important from the point of view of monitoring of technological production processes and can indirectly indicate the type of technology used to produce a drug. That can for example confirm or refute allegations of possible patent infringement (the production technology is usually patented). Raman spectroscopy enables also studying polymorphism of the active compound, due to direct dependence between the molecular structure of compounds and their vibrational spectra. The polymorphism phenomenon is subjected to various research by pharmacists due to a confirmed relationship between the crystal structure of drugs and their bioactivity. In case if the active compound can exist in the anhydrous and hydrated form, its spectral profile enables also its identification.

Spatial distribution of the active compounds in samples of drugs with the help of Raman imaging will be studied in this laboratory exercise. The basis of Raman imaging is described in Chapter 4.4.

Caffeine is a methyl derivative of xanthine found in many various plants such as coffee plant, tea bush, cocoa, yerba mate or guarana berries. This alkaloid is the central nervous system stimulant, showing also weak diuretic properties. It is applied in many analgesics usually in combination with other active substances.

The action of caffeine is multi-faceted and not fully understood. Caffeine is included to the group of stimulants as acting on the central nervous system it decreases fatigue, improves mood and concentration, while negatively influencing long-term memory.

The chemical formula of caffeine is presented in Fig 7.13.1.

Anhydrous caffeine exists in the form of white hexagonal crystals. Hydrated caffeine (monohydrate) forms thin needles.

AIM OF THE EXPERIMENT

1. Identification of the active drug substance using Raman spectroscopy.
2. Performing Raman imaging measurement.

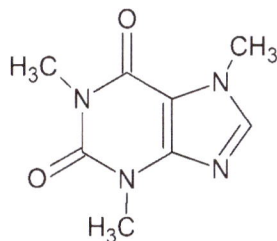

Fig. 7.13.1. The chemical formula of caffeine

3. Learning methods of construction of chemical images showing the distribution of drug components.
4. Interpretation of results obtained in the form of Raman images.

SCIENTIFIC BACKGROUND

1. Theoretical basis of light scattering phenomenon, light scattering mechanism based on virtual energy states; Chapter 2.
2. Construction and principle of operation of a Raman spectrometer; Chapter 2.
3. Basis of Raman imaging; Chapter 4.4.

FURTHER READING

1. Dieing T., Hollricher O., Toporski J. (red.), *Confocal Raman Microscopy*, Springer, Heideberg, 2010, pp. 43-60.
2. M. Barańska, Proniewicz L.M., *Raman mapping of caffeine alkaloid*, Vib. Spectrosc., **48**, 153 (2008).
3. Edwards H.G.M., Farwell D.W., de Oliveira L.F.C., Alia J.-M., Hyaric M.L., de Ameida M.V., *FT-Raman spectroscopic studies of guarana and some extracts*, Anal. Chim. Acta, **532**, 177 (2005).

EQUIPMENT, MATERIALS, CHEMICALS

1. Pure caffeine and analgesics such as Apap-extra, Aspirin Activ, Coffepirine, Coldrex MaxGrip C, Panadol-extra or Scorbolamid.
2. FT Raman spectrometer or dispersive Raman spectrometer with a mapping stage.
3. OPUS, Witec, Cytospec softwares or others for spectral analysis and the construction of Raman images.

PROCEDURE

I. Measurements of Raman spectra and Raman imaging
   1. Record Raman spectrum of pure caffeine in the solid state.
   2. Prepare analgesic samples by removing the drug cover (if applicable), flattening the surface of measured area (if necessary) and position a measured analgesic sample on the scan table.
   3. Record spectrum of the tablet.
   4. Choose measurement parameters (focus, laser power, integration time for a dispersive spectrometer), the size of the measured area and the number

of points in the x and y axis and record a set of Raman spectra taking into account time of measurement and the area of studied sample.

5. Register spectra from the measured area varying sampling density and/or objective magnification.

II. Analysis of Raman imaging results

1. Read from the pure caffeine spectra marker bands of caffeine and assign them to vibrations from Table 7.13.1.

2. If necessary, perform pre-processing of Raman spectra, e.g. baseline correction, smoothing, normalization.

3. Choose marker bands of caffeine and calculate their integral intensity. Make Raman images showing distribution of an active compound based on integration of marker bands.

4. Construct false-colour images using cluster analysis (for instance in Witec Plus software) assuming division for 4,5 and 6 classes and changing the range of wavenumbers chosen to analyse. Based on average spectra of each image choose the number of classes representing the chemical content of an analysed sample.

5. Perform a similar analysis of other measured regions.

REPORT

1. Based on the recorded caffeine spectra, verify if it exists in the anhydrous or hydrated form.

Table 7.13.1. Assignment of the characteristic bands for anhydrous and monohydrated caffeine

Anhydrous caffeine Wavenumber [cm⁻¹]	Caffeine monohydrate Wavenumber [cm⁻¹]	Mode assignment
3121m	3113m	ν(=CH)
2955m	2957s	ν(CH$_3$)
1698s	1698s, 1656m	ν(C=O)
1605s	1600m	ν(C=C)
1551w	1554w	ν(C-C)
1409m	1408	ν(CN)
1361m	1360	ν(CN)
1333s	1328s	ν(CN)
1290m	1284m	ν(CN)
1255m		δ(H-C=N)
1076w	1073w	δ(C-C)
930w	928w	δ(imidazole)
890w		λ(H$_2$O)
647m	644m	δ(O=C-N)
555s	556s	δ(O=C-N)

(s) strong, (m) medium, (w) weak, (ν) streching, (δ) deformation, (λ) libration

2. Present and describe the obtained Raman images showing distribution of caffeine in the studied samples.
3. Show images obtained using cluster analysis. Based on the average spectra, identify characteristic (marker) bands of caffeine and other compounds present in the studied tablets using the composition description of a drug and the reference (literature) spectra.
4. Discuss the usefulness of the Raman imaging and methods used to construct images to study *in situ* the active substances in the sample.

7.14. Characteristics of the cell organelles based on analysis of marker bands and cluster analysis

Katarzyna Majzner

Raman spectroscopy (RS) enables the study of chemical composition of biological samples, including substances such as proteins, nucleic acids, lipids, carbohydrates, heme, carotenoids and inorganic crystals. It is non-invasive and relatively sensitive method for imaging of biochemical changes at the subcellular level.

Measurements of cells in an aqueous environment enables the study of cells at physiological conditions (possibility to study living cells) and tracking of changes within them at subcellular level under the influence of various factors (results of the first Raman measurement of the chromosomes of individual cells was presented in 1990 in the Nature [1]). RS allows *i.a.* monitoring of influence and intracellular distribution of drugs and other bioactive compounds, nanoparticles, as well as non-chemical effects of various factors (e.g. temperature, light, radiation).

The composition of a typical animal cell is shown in the Figure 7.14.1.

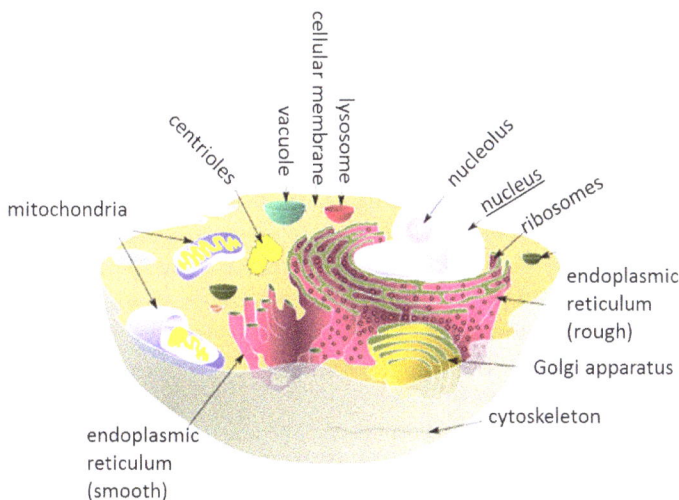

Fig. 7.14.1. The composition of an animal cell [2]

The cells are composed of water (the content can reach up to 70% [3]) and inorganic ions, and organic matter, referred to as dry matter. The proteins, nucleic acids and carbohydrates constitute 80-90% of the dry cell weight, of which over 50% are proteins [3].

7.14.1. The assignment of the major Raman bands in the cell spectra

Individual animal cells, depending on their function, may differ in the structure and chemical composition. This diversity is reflected on the Raman spectra recorded and thus can be the basis of the identification and distinguishing between healthy, pathological and pharmacologically stimulated cells. Despite the differences in the profile of the registered Raman spectra and biochemical diversity of the cells, each spectrum of the cell usually shows certain characteristic features that can be assigned to the appropriate molecular vibrations (Table 7.14.1).

Table 7.14.1. Ranges of Raman bands in spectra of animal cells including an assignment of dominant vibrations, based on [4]

Wavenumber [cm^{-1}]	Mode assignment	Compound assignment
622	C-C out-of-plane bending	proteins
642	C-C out-of-plane bending of Tyr and Phe	proteins
670	ring breathing of G	DNA
725	ring breathing of A	nucleic acids
746	ring breathing of T	nucleic acids
767	pyrimidine ring breathing; ring breathing of Trp	nucleic acids, proteins
785-790	ring breathing of U, T, C; backbone O-P-O	nucleic acids
816	Pro, Hyp, Tyr, PO$_2^-$ stretching	proteins, nucleic acids, (one of the two distinct bands for RNA – a second is at 1240 cm^{-1})
858-861	C-C stretching	Pro, Hyp, Tyr
884	CH$_2$ deformation mode (rocking)	proteins
944	skeletal modes	polysaccharides
1008	ring breathing	Phe
1031-1035	C-H bending	Phe
1034-1038	C-N stretching	proteins
1085-1100	PO$_2^-$ symmetric stretching	nucleic acids, phospholipids
1130-1135	C-C skeletal of acyl backbone in lipid (*trans* conformation)	lipids, phospholipids
1184	C, G, A	nucleic acids
1244	asymmetric stretching PO$_2^-$, C-N stretching, modes of pyrimidine bases (C, T)	nucleic acids

Table 7.14.1. (continued)

1268	amide III; C-N stretching in α-helix	proteins
1304	CH_2/CH_3 deformation	lipids, fatty acids
1318	ring breathing of G, CH deformation, amide III (α-helix)	nucleic acids, proteins
1339-1342	CH_2/CH_3 wagging, twisting and/or bending of lipids, CH deformation (proteins and carbohydrates), G modes	nucleic acids, proteins, lipids
1449-1455	CH_2/CH_3 bending in-plane	proteins, lipids
1580-1585	pyrimidine ring; C=C bending in Phe	nucleic acids, heme proteins
1654	C=C stretching	triacylglycerols
1658-1670	amide I, C=C stretching	proteins, lipids,
1660-1665	C=C stretching; amide I α-helix)	mainly proteins
1670	C=C stretching of steroid ring and *trans*, amide I (anti-parallel β-sheet)	proteins, lipids, fatty acids
1740	C=O stretching	esters, lipids, phospholipids
2850-2870	CH_2 symmetric stretching	lipids, fatty acids
2885	Fermi resonance of CH_2 stretching; symmetric stretching CH_3; asymmetric stretching CH_2	lipids, fatty acids, proteins
2933-2945	asymmetric stretching CH_2	lipids, fatty acids, proteins
3000-3030	=CH stretching	lipids, fatty acids

Most information about the biochemical composition of biological samples can be found in the fingerprint spectral range: 0-1800 cm⁻¹. The presence of proteins manifests itself by two characteristic Raman bands referred to as amide I and amide III. Since the first of these is derived from the stretching vibration of C=O, it provides information about the secondary structure of proteins. Information on content and structure of the protein in the cells can be obtained from the analysis of bands at 1660 cm⁻¹ (amide I), 1450 cm⁻¹ (CH_2 bending), 1100-1375 cm⁻¹ (amide III) and 1004-1008 cm⁻¹ (phenylalanine). The lipid composition of cells can be determined based on the bands: 1740 cm⁻¹ (C=O); 1656 cm⁻¹ (*cis* C=C); 1441 cm⁻¹ (CH_2 bending) and 1304 cm⁻¹ (CH_2 twisting). Not without a significance is density of the investigated area of the sample because the area of reduced density, cytoplasm for instance, is manifested by a lower signal to noise ratio – especially in the fingerprint range.

The range of high wavenumbers (2800-3030 cm⁻¹) provides information on the protein-lipid composition. It is quite difficult to interpret that spectral range, but it is very useful in the analysis of lipids due to band at 3005-3015 cm⁻¹ (stretching vibration of =C-H). The intensity of this band can be correlated to the amount of double bonds in the molecule of lipid/fatty acid. The spectral difference associated with lipids can be tracked by analyzing the band within a range of 2850 – 2900 cm⁻¹, which is more prominent for the areas of the cytoplasm around the nucleus (the area of the endoplasmic reticulum and mitochondria).

7.14.2. An analysis of cell Raman imaging results

Raman imaging leads to large amounts of spectral and spatial data. For their analysis a standard approach as well as chemometric methods are used, *i.e.* analysis of marker bands and cluster analysis.

Standard approach, namely analysis of chemical composition based on the analysis of the marker bands present in the spectrum, allows for extraction only a part of information about the complex systems, which are biological samples. The required condition for standard approach in data analysis is the presence of the marker band in the spectra and the possibility of their identification.

Cluster analysis is a multivariate analysis combined with a clustering of pixels, based on the similarity of their spectral properties, which facilitates the analysis of the obtained results and allows the use of spatial information.

With detailed knowledge about the position of each Raman band in the spectrum the assignment to different biochemical compounds (such as lipids, proteins or nucleic acids) is relatively easy. Consequently, the spectrum can be assign to specific cellular substructures. An example of characteristic Raman spectra taken from the cell with reference to the specific organelles, based on the presence of the marker bands, is shown in the Figure 7.14.2. The presented spectra were averaged from individual classes obtained from Hierarchical Cluster Analysis (HCA is discussed in chapter 5).

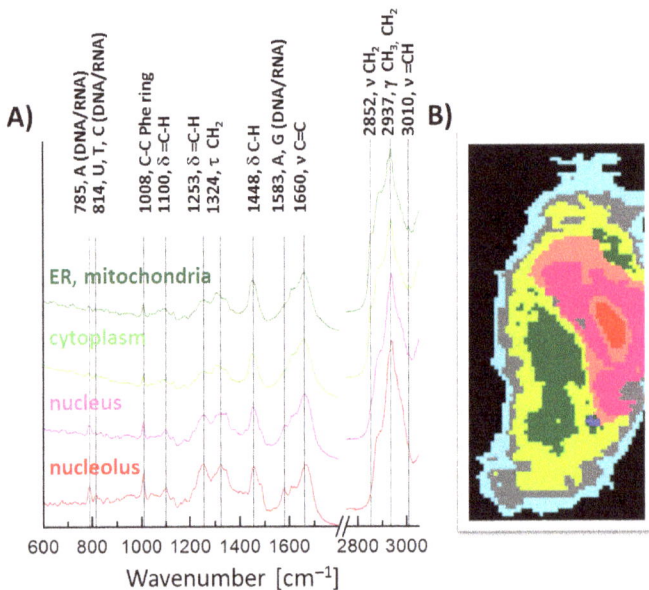

Fig. 7.14.2. Averaged spectra of selected classes (Hierarchical Cluster Analysis) assigned to specific cellular structures with the distribution of these structures on the Raman map (endothelial cell line EA.hy926). Characteristic bands of the cell in Raman spectra with corresponding vibrations. Measurement parameters: 532 nm laser line, the integration time of 0.5s, the size of the imaged area of 16.7×34.1 µm², the spatial resolution 0.32 µm (map size 52×106 points)

The most diverse Raman spectra originate from such structures as nucleus, nucleolus, endoplasmic reticulum and cytoplasm. Structures smaller, due to the limited spatial resolution cannot be identified and visualized (e.g. cell membrane, single mitochondria, lysosomes, or ribosomes). It is worth noting that various animal cells due to other functions in the body can vary in size and composition of intracellular structures, e.g. number and size of mitochondria or the presence of lipid bodies/droplets.

AIM OF THE EXPERIMENT

1. Measurement of cells using Raman imaging technique.
2. An analysis of chemical composition of the cell and the identification of intracellular structures based on:
   a) analysis of selected marker bands,
   b) cluster analysis.

SCIENTIFIC BACKGROUND

1. Fundamentals and theory of Raman spectroscopy and Raman imaging; Chapters 2 and 4.4.
2. Construction and principles of operating the Raman spectrometer and confocal microscope; Chapters 2 and 4.4.
3. Basics of chemometric methods used to analysis of the Raman spectra.
4. Basics of animal cell structure and its biochemical composition.

EQUIPMENT, MATERIALS AND CHEMICALS

1. A sample in the form of fixed animal cells obtained from *in vitro* culture seeded on the substrate allowing Raman measurement (e.g. CaF_2).
2. Raman spectrometer adapted for imaging.
3. Data analysis software with the ability to perform chemometric analysis of spectra obtained by Raman imaging, e.g. CytoSpec.

PROCEDURE

1. Carry out the measurement of Raman imaging of animal cells selecting carefully measurement conditions (laser power, the measurement time) to avoid damage of the sample, and to get a good signal to noise ratio within the shortest time.
2. Select and indicate marker bands in the Raman spectra of the main cell components.
3. Carry out the analysis of selected, by the student, marker bands. Analyze the obtained results by discussing the distribution of selected components and assign them to the animal cell intracellular structures.
4. Carry out the cluster analysis using selected method/methods. Analyze the obtained results by discussing individual classes and assign them to the animal cell intracellular structures based on the profile of the averaged spectra and the location of obtained classes.

REPORT

1. On the base of the results obtained by applying analysis of the marker bands, discuss the distribution of the main components of animal cells.
2. On the base of the average spectra from the cluster analysis, indicate the marker bands of the main intracellular structures. Discuss spectral differences and similarities between subcellular structures.
3. Find Raman bands positions from the average spectra of the animal cell and assign them to vibrations based on reference spectra found in the literature. Assign obtained classes to intracellular structures.
4. Discuss the utility of Raman imaging technique to *in vitro* studies and usefulness of chemometric methods for spectral data mining.

References

1. Puppels G.J., de Mul F.F., Otto C., Greve J., Robert-Nicoud M., Arndt-Jovin D.J., Jovin T.M., *Studying single living cells and chromosomes by confocal Raman microspectroscopy*, Nature, **347**, 301 (1990).
2. Petibois C., *Imaging methods for elemental, chemical, molecular, and morphological analyses of single cells*, Anal. Bioanal. Chem., **397**, 2051 (2010).
3. Cooper G., *The Molecular Composition of Cells*, Cell A. Mol. Approach., 2nd Ed. (2000).
4. Movasaghi Z., Rehman S., Rehman I.U., *Raman Spectroscopy of Biological Tissues*, Appl. Spectrosc. Rev., **42**, 493 (2007).
5. Hedegaard M., Matthäus C., Hassing S., Krafft C., Diem M., Popp J., *Spectral unmixing and clustering algorithms for assessment of single cells by Raman microscopic imaging*, Theor. Chem. Acc., **130**, 1249 (2011).

7.15. *In vitro* and *in vivo* Raman imaging of unicellular carotenoid producers

Marta Z. Pacia, Agnieszka Kaczor

Xanthophylls are a group of carotenoid pigments containing oxygen in theirs structure. The primary xanthophylls (e.g. zeaxanthin and lutein; Fig. 7.15.1 A-B) can be produced by higher plants, while secondary xanthophylls (e.g. fucoxanthin and astaxanthin; Fig. 7.15.1C-D) are synthesized *de novo* only by unicellular organisms [1]. Xanthophylls produced by algae (e.g. *Haematococcus pluvialis*) and yeasts (*Phaffia rhodozyma, Rhodotorula mucilaginosa*) have very strong antioxidant properties and naturally protect cells against free radicals that help to maintain cellular redox equilibrium.

Unicellular algae *Haematococcus pluvialis* or yeast *Phaffia rhodozyma* have the ability to produce astaxanthin, especially under the stressful environmental factors.[2] The structure of astaxanthin in the algae cells as well as the change of its structure upon thermal stress have been investigated by Raman spectroscopy [3, 4].

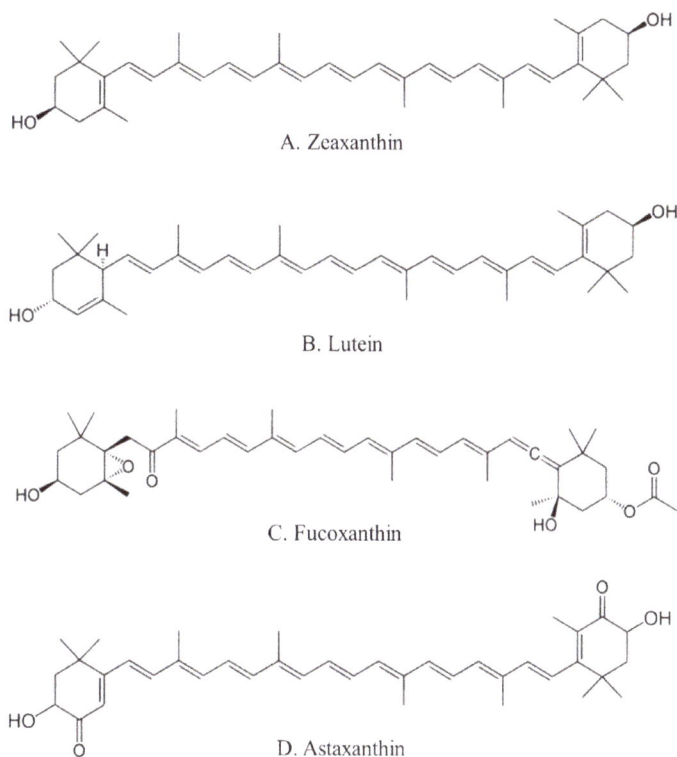

Fig. 7.15.1. Structures of chosen xantophylls

R. mucilaginosa can produce four major carotenoid pigments: β-, and γ-carotene, torulene and torularhodin that are responsible for orange-red color of yeast cells. Torularhodin is the major yeast carotenoid component (up to 80% of the total carotenoid pigments), however the carotenoid content depends strongly on the yeast strain and the development phase [5].

Raman spectroscopy enables investigation of unnicellular organisms both *in vitro* (fixed cells) [6] and *in vivo* (live cells) [7]. The *in vivo* measurements are more challenging and require maintaining the appropriate measurement conditions.

7.15.1. Spectral characteristics of cells compartments

Carotenoids

Raman spectroscopy is a particularly convenient method to study carotenoid due to the long chain of alternating double and single bonds that is responsible for unusually high Raman scattering cross-section associated with chain vibrations (particularly C=C and C-C stretching vibrations). Therefore, Raman imaging due to very high spatial resolution (up to *ca.* 360 nm at λ=532 nm excitation) provides information about carotenoids distribution in the single cells even at low local concentration of these compounds. The exemplary spectrum of carotenoid in the *R. mucilaginosa* yeast cell is presented in Figure 7.15.2.

The carotenoids can be easily visualized by integration of the major carotenoid bands: at 1512 cm^{-1} (v_1) and 1156 cm^{-1} (v_2), assigned to the C=C and C–C stretching modes, respectively. Additionally, in-plane rocking modes of CH$_3$ groups attached to the polyene chain can be observed as a medium intense band at 1003 cm^{-1}. The position of v_1 band depends on various factors such as the number of conjugated double bonds in the carotenoid structure, binding with proteins or lipids, and the conformation of the carotenoid molecules [8].

Lipids

Raman imaging enables studying distribution of lipid bodies and their composition in the single algal and yeast cells. The lipid bodies in yeast are not only the storage of lipids but they can also act as signaling structures [9]. Lipids can be responsible for removal of excess fatty acid from the cell and, conversely, provide fatty acids for the synthesis of membrane phospholipids. Lipid bodies observed in *R. mucilaginosa* cells (Fig. 7.15.3) occurred to be rich in unsaturated triacylglycerols.

Based on Raman spectra of lipids, the information about degree of unsaturation can be elucidated [10]. The ratio of the band due to the C=C stretching

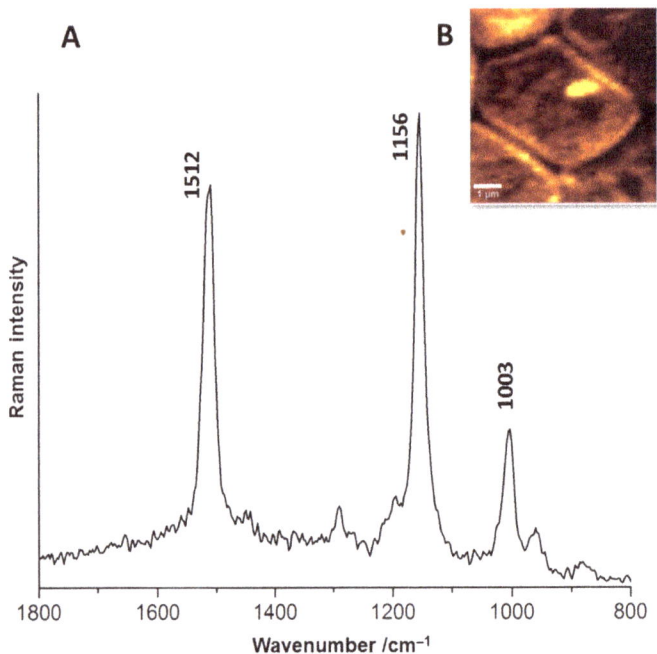

Fig. 7.15.2. Raman spectrum of carotenoid observed in *R. muciliganosa* cells (A) and the representative Raman image of distribution of carotenoids (integration over the band at 1156 cm^{-1}, B)

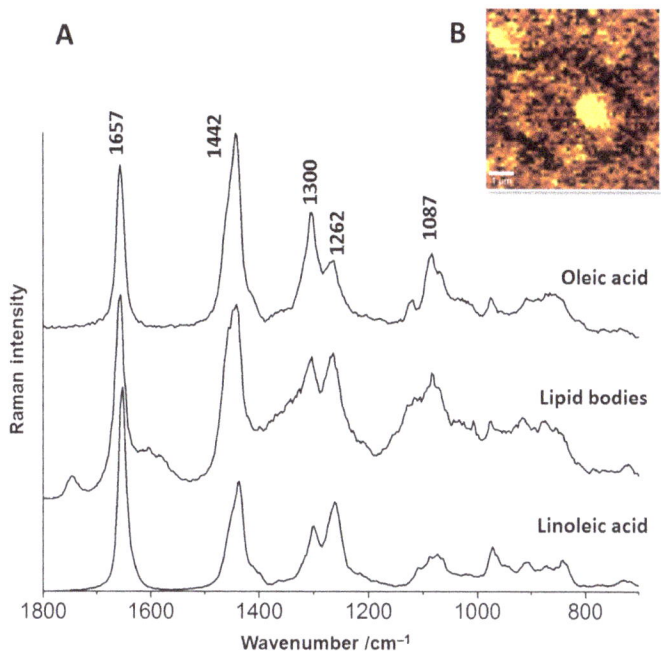

Fig 7.15.3. The Raman spectrum of a lipid body observed in *R. muciliganosa* cells compared with two standards: oleic and linoleic acids (A). The representative Raman image of distribution of lipid bodies (integration over the band at 2857 cm⁻¹, B)

vibration (1657 cm⁻¹) to that related to the CH_2 scissoring mode (1442 cm⁻¹) as well as the ratio of the band assigned to the CH_2 twisting mode (1300 cm⁻¹) to the feature due to the =C-H in plane deformation (1262 cm⁻¹) could be used for reliable determination of the total degree of unsaturation [11]. According to the data presented in Figure 7.15.3, it can be concluded that a degree of unsaturation of lipids in lipid bodies of *R. mucilaginosa* cells is between oleic acid (one double bond) and linoleic acid (two double bonds).

Hemoproteins

The aggregations of hemoproteins can be also observed in the unicellular yeast cells.[12] Using Raman spectroscopy, particularly upon green laser line excitation (514 or 532 nm), the convenient detection of the hemoproteins signal is possible due to the resonance enhancement of the prostatic heme group signals (three high intense bands located at 1588, 1130 and 750 cm⁻¹). These bands as well as the increase of the background signal are characteristic Raman markers of hemoproteins.[13]

Cell wall

The cell wall of yeast cells is composed of multicomponent layers. Major structural constituents of the cell wall are polysaccharides (80–90%), mainly glucans and mannans, with a minor percentage of chitin.

Generally, the cell wall skeleton is based on the *1,3*-glucan with chitin complex while *1,6*-glucans are responsible for the connection between the inner and outer wall components. Mannans combined with proteins form the outermost layer of the cell wall. Such structure of the cell wall cerates the effective conditions for its protective and building functions [14]. The Raman spectrum of the cell wall of *R. mucilaginosa* (Fig. 7.15.4) is dominated by the following bands: 1460 cm^{-1} (the CH$_2$ deformation vibration), 1375 cm^{-1} (the C-C-H deformation vibration), 1266 cm^{-1} (the =C-H deformation vibration), 1082 cm^{-1} (the C-C stretching vibration) and 443 cm^{-1} (the ring deformation modes) [15].

AIM OF THE EXPERIMENT

The aim of the experiment is the application of high-resolution Raman imaging for detection of the subcellular composition of unicellular carotenoids producers and the analysis of the obtained data for investigation of the subcellular structures.

SCIENTIFIC BACKGROUND

1. Fundamentals of Raman spectroscopy; Chapter 2.

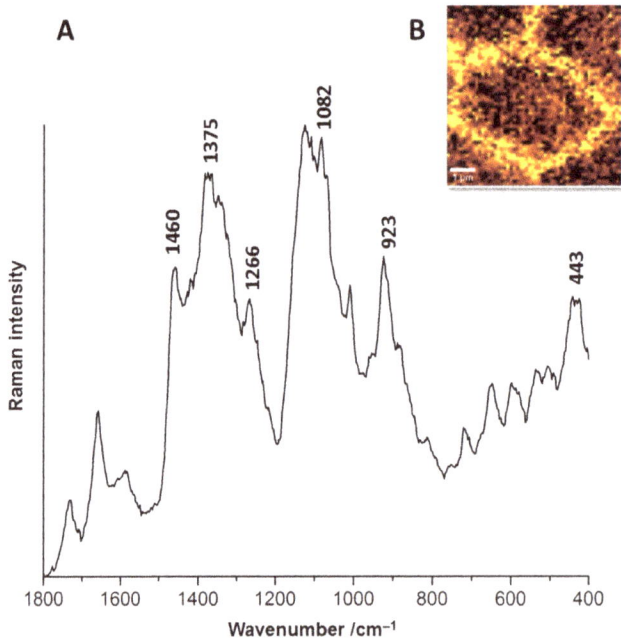

Fig. 7.15.4. The Raman spectrum of cell wall observed in *R. muciliganosa* cells (A) and the representative image showing of the shape and localization of the cell wall (integration over the band at 1460 cm^{-1}, B)

2. Basic knowledge about Raman imaging; Chapter 4.4.
3. Construction and principle of operation of dispersive Raman spectrometer; Chapter 2.
4. Spatial resolution – the Rayleigh criterion; Chapter 4.4.
5. Basic information about biochemical composition of subcellular components of yeast/algae cells and their spectral characteristics.

FURTHER READING

1. Kaczor A., Barańska M., *Structural changes of carotenoid astaxanthin in a single algal cell monitored in situ by Raman spectroscopy*, Anal. Chem., **83**, 7763 (2011).
2. Kaczor A., Turnau K., Barańska M., *In situ Raman imaging of astaxanthin in a single microalgal cell*, Analyst. **136**, 1109 (2011).
3. Schulz H., Barańska M., Baranski R., *Potential of NIR-FT-Raman spectroscopy in natural carotenoid analysis*, Biopolymers. **77**, 212 (2005).
4. Schulz H., Barańska M., *Identification and quantification of valuable plant substances by IR and Raman spectroscopy*, Vibr. Spectr., **43**, 13 (2007).
5. Malamud D., Borralho L., Panek A., Mattoon J., *Modulation of cytochrome biosynthesis in yeast by antimetabolite action of levulinic acid*, J. Bacter., **138**, 799 (1979).
6. Ryguła A., Majzner K., Marzec K., Kaczor A., Pilarczyk M., Barańska M., *Raman spectroscopy of proteins: a review*, J. Raman Spectr., **44**, 1061 (2013).

EQUIPMENT, MATERIALS, CHEMICALS

1. Raman spectrometer with an excitation in Vis region, e.g. WITec alpha 300.
2. Samples of unicellular organisms e.g. yeast or algae.
3. Sodium chloride 0.9% solution.
4. Microscope slides, coverslip glass, calcium fluoride slides.

PROCEDURE

1. Sample preparation:
   - *In vitro*: place the sample of the unicellular organism on the calcium fluoride slide, perform a smear and wait for a sample to dry.
   - *In vivo*: place the sample of the unicellular organism on the calcium fluoride slide, add few drops of sodium chloride solution, cover the sample with coverslip glass and protect the effluence of the sodium chloride solution using a quick-drying lacquer.
4. Place the sample under a microscope and focus on the cell surface.
5. Select the measurement conditions (the size of the image, the number of points, the number of scans, integration time, laser power).
6. Start the scan using air objective with e.g. 100x magnification for *in vitro* or oil objective with 60× magnification for *in vivo* measurements. If it is necessary, repeat the measurements by choosing the same place of sample and optimize the measurement conditions.

REPORT

1. Find a proper marker band for specific compartments of unicellular organisms (yeast/algae) and integrate its area to visualize the size and distribution the subcellular components.
2. Extract the single spectra of the individual components of the yeast/algae using the Project WITec 2.10 program.
3. Present Raman images for different components of unicellular cells together with the spectra with the band assignments. Comment on the resulting images, discuss the distribution of the individual cellular components, their size, etc.
4. Discuss the usefulness of Raman spectroscopy to determine the subcellular composition of unicellular organisms and present possible practical applications of the method to study yeast/algae cells.

References

1. Katayama T., Yokoyama H., Chichester C., *The biosynthesis of astaxanthin. I. The structure of α-doradexanthin and β-doradexanthin*, Int. J. Biochem., **1**, 438 (1970).
2. Boussiba S., *Carotenogenesis in the green alga Haematococcus pluvialis: cellular physiology and stress response*, Physiologia Plantarum, **108**, 111 (2000).
3. Kaczor A., Barańska M., *Structural changes of carotenoid astaxanthin in a single algal cell monitored in situ by Raman spectroscopy*, Anal. Chem., **83**, 7763 (2011).
4. Kaczor A., Turnau K., Barańska M., *In situ Raman imaging of astaxanthin in a single microalgal cell*, Analyst. **136**, 1109 (2011).
5. Moliné M., Flores M.R., Libkind D., del Carmen Diéguez M., Farías M.E., van Broock M., *Photoprotection by carotenoid pigments in the yeast Rhodotorula mucilaginosa: the role of torularhodin*, Photochem. PhotobiolSciences, **9**, 1145 (2010).
6. Rösch P., Schmitt M., Kiefer W., Popp J., *The identification of microorganisms by micro--Raman spectroscopy*, J. Mol. Structr., **661**, 363 (2003).
7. Collins A.M., Jones H.D., Han D., Hu Q., Beechem T.E., Timlin J.A., *Carotenoid distribution in living cells of Haematococcus pluvialis (Chlorophyceae)*, PloS one, **6**, e24302 (2011).
8. Schulz H., Barańska M., Baranski R., *Potential of NIR-FT-Raman spectroscopy in natural carotenoid analysis*, Biopolymers, **77**, 212 (2005).
9. Czabany T., Athenstaedt K., Daum G., *Synthesis, storage and degradation of neutral lipids in yeast*, Biochim. Biophys. Acta (BBA), **1771**, 299 (2007).
10. Krafft C., Neudert L., Simat T., Salzer R., *Near infrared Raman spectra of human brain lipids*, Spectrochim. Acta Part A: Mol and Biomol Spectr., **61**, 1529 (2005).
11. Schulz H., Barańska M., *Identification and quantification of valuable plant substances by IR and Raman spectroscopy*, Vib. Spectrosc., **43**, 13 (2007).
12. Malamud D., Borralho L., Panek A., Mattoon J., *Modulation of cytochrome biosynthesis in yeast by antimetabolite action of levulinic acid*, J. Bacter., **138**, 799 (1979).
13. Ryguła A., Majzner K., Marzec K., Kaczor A., Pilarczyk M., Barańska M., *Raman spectroscopy of proteins: a review*, J Raman Spectr., **44**, 1061 (2013).
14. Lipke P.N., Ovalle R., *Cell wall architecture in yeast: new structure and new challenges*, J. Bacter., **180**, 3735 (1998).
15. Novák M., Synytsya A., Gedeon O., Slepička P., Procházka V., Synytsya A., et al. *Yeast β (1-3),(1-6)-d-glucan films: Preparation and characterization of some structural and physical properties*, Carbohydr. Polymers, **87**, 2496 (2012).

List of figures

by integration of IR bands specific for components of the film. Illustration of courtesy of Dr. M. Kansiz and Agilent Technologies

is eliminated)

List of tables

www.ingramcontent.com/pod-product-compliance
Lightning Source LLC
Chambersburg PA
CBHW061410210326

41598CB00035B/6161